Animal Constructions and Technological Knowledge

Postphenomenology and the Philosophy of Technology

Technological advances affect everything from our understandings of ethics, politics, and communication, to gender, science, and selfhood. Philosophical reflection on technology helps draw out and analyze the nature of these changes, and helps us understand both the broad patterns and the concrete details of technological effects. This book series provides a publication outlet for the field of the philosophy of technology in general, and the school of thought called "postphenomenology" in particular. Philosophy of technology applies insights from the history of philosophy to current issues in technology, and reflects on how technological developments change our understanding of philosophical issues. In response, postphenomenology analyzes human relationships with technologies, while integrating philosophical commitments of the American pragmatist tradition of thought.

Animal Constructions and Technical Knowledge, by Ashley Shew
Using Knowledge: On the Rationality of Science, Technology, and Medicine, by Ingemar Nordin
Postphenomenology and Media: Essays on Human–Media–World Relations, edited by Yoni Van Den Eede, Stacey O. Irwin, and Galit Wellner
Diphtheria Serum as a Technological Object: A Philosophical Analysis of Serotherapy in France 1894–1900, by Jonathan Simon
Digital Media: Human-Technology Connection, by Stacey O. Irwin
Acoustic Technics, by Don Ihde
A Postphenomenological Inquiry of Cell Phones: Genealogies, Meanings, and Becoming, by Galit P. Wellner
Technoscience and Postphenomenology: The Manhattan Papers, edited by Jan Kyrre Berg O. Friis and Robert P. Crease
Postphenomenological Investigations: Essays on Human–Technology Relations, edited by Peter-Paul Verbeek and Robert Rosenberger
Design, Meditation, and the Posthuman, edited by Dennis M. Weiss, Amy D. Propen, and Colbey Emmerson Reid

Animal Constructions and Technological Knowledge

Ashley Shew

LEXINGTON BOOKS
Lanham • Boulder • New York • London

Published by Lexington Books
An imprint of The Rowman & Littlefield Publishing Group, Inc.
4501 Forbes Boulevard, Suite 200, Lanham, Maryland 20706
www.rowman.com

Unit A, Whitacre Mews, 26-34 Stannary Street, London SE11 4AB

British Library Cataloguing in Publication Information Available

Library of Congress Cataloging-in-Publication Data

Names: Shew, Ashley, 1983- author.
Title: Animal constructions and technological knowledge / Ashley Shew.
Description: Lanham : Lexington Books, 2017. | Series: Postphenomenology and the philosophy of technology | Includes bibliographical references and index.
Identifiers: LCCN 2017036567 (print) | LCCN 2017038549 (ebook) | ISBN 9781498543125 (Electronic) | ISBN 9781498543118 (cloth : alk. paper)
Subjects: LCSH: Tool use in animals. | Technology--Philosophy. | Animal intelligence. | Animal behavior.
Classification: LCC QL785 (ebook) | LCC QL785 .S446 2017 (print) | DDC 591.5/13--dc23
LC record available at https://lccn.loc.gov/2017036567

∞™ The paper used in this publication meets the minimum requirements of American National Standard for Information Sciences Permanence of Paper for Printed Library Materials, ANSI/NISO Z39.48-1992.

Printed in the United States of America

for Zora, human-Leah, and gorilla-Leah

Contents

ONE

Introduction

Recently, a video showing an orangutan making a hammock out of a sheet left in her cage went viral on social media.[1] Accounts of animal intelligence are now entering the mainstream; news stories and Internet videos show crows, apes, and octopuses using tools or solving problems. There is a growing public awareness that animals are sensitive and intelligent, and this public awareness parallels a recent growth in academic studies of animal tool use. Animal ethics has become a robust field of study; legislation governing animal use in research, while imperfect, makes distinctions between species and types of animals. Compared to these advances in animal ethics, scientific studies of animal tool use, and the popular understanding and anecdotal accounts of animal intelligence, the philosophical literature connecting animals' tool use to existing philosophical theories of technology and intelligence lags behind. This book bridges that gap, linking developments in our knowledge about animal intelligence and behavior to our epistemologies of technology and the human and offering a model for incorporating non-humans into our accounts of the world.

This book develops an epistemology of technologies that considers some of the actions and products of non-human animals as technological in character. Typically, we think of things as "technological" only when they are made, designed, employed, or implemented by human beings, and, in fact, the idea of technological knowledge—knowledge of how things work, of how to work things, and of how to create devices—has generally been discussed by historians, anthropologists, and philosophers of technology and of engineering in terms of *human* agency. Though technological knowledge is occasionally linked to computers and artificial intelligence, it is usually viewed as being the domain of engineers and craftspeople. This assignment of technological know-how to humans alone, however, neglects the technological knowledge and know-how implemented by non-human animals.

The suggestion that non-human animals have "technological" anything may strike some readers as absurd, precisely because we define technology in terms of humanity. But if we disregard these definitions' "human clause" (which I argue we ought to do), some animals' ability to shape and employ tools fits neatly into the definitions of technological knowledge produced by philosophers of technology. In this project, I advance a more inclusive account of technology and tool use, give an argument for "technological knowledge" as including animal tool-making and tool use, and look at actual cases of tool use in non-human animals. This discussion also necessarily touches on the content and nature of animal minds,[2] because technological projects are projects that are intentional, designed, and purposeful; to classify animal productions as technological, we need to know something about whether animals are capable of intentionality.

This reconsideration of animal tool use within the framework of philosophy of technology grows out of the wealth of observational data collected about animal minds and animal behaviors in the past three decades. There is a growing literature indicating that animal tool use is much more sophisticated than philosophers had previously supposed. New observations have led to an impressive catalog of animal tools and their uses: dolphins using sponges to help rustle up food, a chimp arming himself with rocks to throw at zoo-goers, and crows that shape twigs in a number of distinct ways for different purposes. Researchers have developed new observational techniques such as camera traps, which have contributed to some of these revelations, but cameras in the hands of amateurs (for example, an amateur YouTube video of birds stealing bags of chips from a convenience store) have also contributed to the increasing recognition that non-human animals can manipulate their environments in very significant ways. And new techniques developed by animals—the birds in the YouTube video open the bags of Doritos they stole from the convenience store (CorellianScoundrel 2008)—are not the products of instinct, in any pure sense.

In these videos and observational accounts, and in the made objects examined by researchers, we see animals using patterns, techniques, and strategies; transmitting know-how; and mimicking other animals' behaviors—processes similar to the technological behaviors engaged in by human beings. Some readers may object, believing that no matter how sophisticated they seem, animal behaviors and artefacts are evidence of instinctual responses to a stimulus, not examples of planning or goal-directed problem-solving. This book will explore some contrast cases—animal-made creations that we might not count as technologies, such as spider webs and beaver dams. But, as we will see, many cases of animal artifact-use evince the same qualities that define human technology use:

planning, problem-solving, design, and innovation. In this book, my guiding question is this: *Do animal behaviors and constructions count as technological knowledge in the same way that human enterprises do?*[3] I arrive at an account of technological behavior that is inclusive of several approaches to technological knowledge and is sensitive to both intentional and instinctive types of animal cases.

In investigating this question, I look at the many ways in which technological knowledge has been described and test each element against some animal case study or studies; as I will show, each element of technological knowledge can be applied as well to non-human tool-users as to their human counterparts. With this book, I aim to both establish that animal tool use should be investigated in epistemologies of technology and initiate this work. The theoretical framework with which I begin this project unites two different discourses on technological knowledge, one on the knowledge content of objects and the other on the learned skill, or know-how, of the tool-user or tool-maker. By treating these two discourses as axes of interest, and by mapping animal constructions alongside human technologies on these two axes, this book unites dialogues about biological and engineering design and provides a more coherent, unified account of made things.

BACKGROUND: TECHNOLOGICAL KNOWLEDGE

This project draws primarily from three types of literature: work on technological knowledge by philosophers of technology, work from philosophers of biology that depend on animal studies or thoughts about animal minds, and work on animal behavior and cognition from biologists, zookeepers, hobbyists, and animal psychologists.

There are two major types of accounts of technological knowledge: one type sees knowledge as knowledge that is enacted upon the world through material devices, and the other type sees technical knowledge as being encapsulated in artefacts. The first type of account has often positioned technological knowledge as being in tension with scientific or theoretical knowledge; this strain of thought represents a negative reaction to the idea that technology is the handmaiden of science, that technology is simply "applied science." These accounts, developed by theorists such as A. Rupert Hall (1978), Edwin T. Layton, Jr. (1974), Walter Vincenti (1990), Carl Mitcham (1994), and Joseph C. Pitt (2000), aim to describe a domain of knowledge that is specific to technology. This first type of account rests upon the distinction between "knowing how" and "knowing that"; technological knowledge is knowledge about making or doing something. Theorists of the first type of account differ from one another in small ways (they flesh out their concepts using different subcategories and types, and some disagree about where to draw the line between know-how and know-that: for example, Vincenti thinks both are involved in engineering knowledge, and Hall thinks they are more closely connected than Layton does), but these accounts are unified by their shared identification of the special character of knowledge produced and used by human beings in their technological endeavors.

This first type of account nearly universally insists that human beings are the sole possessors of technical knowledge. For example, A. Rupert Hall's account of technology defines people as makers of things (*homo faber*). He equates the history of technology with human history, connecting a young boy's knowledge of bicycle riding to prehistoric skills of blade-forging and clay-pot-making. Using the notion of know-how, which is fundamental to many definitions of technological knowledge, Hall explains that "knowing how to ride a bicycle consists in acquiring a certain set of neuro-muscular co-ordinations as does mastery of many technical skills such as throwing a clay pot, making a good saw cut and ploughing a straight furrow" (1978, 94).

"Knowing how" involves coordination of the body, the learning of a skill. Hall makes a further distinction between "age-old" know-how and the mathematicized knowledge that underpins steam engines and airplanes, noting that the character of technological knowledge has changed over time, in part through application of science and in part through imitation of science (98). Though Hall describes technology in terms of what man has made, his discussion of know-how leaves room for animal cases to be analyzed on his terms; there is no reason that non-human actors could not perform the neuro-muscular coordination and skill that Hall offers as the basis of technological knowledge. However, Hall does offer a warning about the reading of ancient artifacts that might be equally relevant to attempts to apply his theory of technological knowledge to non-human actors: we must be careful not to posit too much about the cognitive capabilities of those who produced such artifacts.

Edwin Layton offers a different account of technical knowledge—one that Hall disagrees with. Layton sees science and technology as wholly different bodies of knowledge, which feed off each other but are not as interconnected as they are in Hall's model. In two different articles, "Mirror-Image Twins: The Communities of Science and Technology in 19th-Century America" (1971) and "Through the Looking Glass, or News from Lake Mirror Image" (1987), Layton depicts science and technology as mirror images of one another: they are very much alike, but not exactly so, and they are separate from one another. Layton does not see science as passing responsibility for practical, applied knowledge down to technology; instead, he argues that the flows of knowledge from technology to science are as symmetric to those from science to technology. Technological knowledge, for Layton, can be divided into three parts: technological science, or theory; design; and technique (1987, 604). The field of design involves all three parts, and is "knowing how" (a phrase that

Layton uses as a synonym for technological knowledge) "at the highest level" (ibid.). Layton looks to engineering design for his historical examples in all cases.

Walter Vincenti, in his 1990 *What Engineers Know and How They Know It*, contributes to the study of technological knowledge by using aeronautical engineering case studies to explain the design process; although he acknowledges that his ideas are not necessarily generalizable to the entire domain of technological knowledge (1990, 7), the use of engineering design case studies has become common in discussions of technological knowledge. This focus on engineering makes sense; though people in many fields design and use technologies, this process is at the core of what engineers do. While this engineering-led approach helps us clarify the design process and recognize knowledge transformations across many steps in the creation of technologies, it tells us little about what constitutes technological knowledge *in general*, rather than what constitutes technical knowledge *for engineers*. As historian-philosopher of technology Ann Johnson reminds us, too much focus on the discipline of engineering may limit our understanding of technology more broadly:

> Fixing a notion of technology as knowledge runs the risk that some technologies or some knowledge forms would be excluded or eliminated, as has been the case in defining scientific knowledge. Scholars in the technology as knowledge tradition have carefully avoided limiting definitions of technological knowledge in an explicit effort to avoid some of the restrictions that have arisen through the epistemology of science. (Johnson 2005, 555)

Despite a disciplinary commitment to flexibility in definitions of technological knowledge, the study of the epistemology of technology has become largely focused on engineering practice, infrastructure, and community.

Engineering case studies feature prominently in discussions of technological knowledge in the work of Layton, Vincenti, Mitcham, Pitt, Johnson, and many others. While these engineering cases are useful in illustrating the process of developing and employing technological knowledge and in clarifying the difference between scientific knowledge and technological knowledge (a move that was once very important to philosophers of technology), there is a large body of technological knowledge outside the realm of formal professional engineering. In a Venn diagram, engineering knowledge would be a small circle completely located inside the larger circle of technological knowledge: all engineering knowledge is technological knowledge, but not all technological knowledge is engineering knowledge. By looking at technological knowledge only in terms of what engineers do, we severely constrain our discussion, and the engineering-focused approach forecloses the notion that animals might use and deploy technological knowledge.

Joseph C. Pitt (2000) extends findings from studies of engineering knowledge to technological knowledge more generally, but his very definition of technology, "humanity at work," is based on the "human clause." Pitt believes that his definition could change to include "the activity of beavers or aliens," but that first we would need to have "a good idea of purposeful activity for non-humans" (11), and we need an account of purposeful activity for humans before trying to produce one for animals. Pitt, in a model for his account of technological knowledge, explains that "humans at work" engage in an input–output transformation process in which decisions and deliberations about knowledge are fed back into the system, and that these evaluation feedback loops allow for evaluation of progress. While I cannot (yet) investigate aliens at work, the activity of, say, beavers may warrant consideration under Pitt's model—at least as a contrast to the sorts of knowledge Pitt is interested in.

In contrast to these accounts of the first type, which see technological knowledge as being a kind of applied science—knowledge that is simply enacted on the world via devices—the second type of account, which grows out of work in the philosophy of science, sees knowledge as embodied in the objects themselves. In this second type of account, technological knowledge is a form of Davis Baird's "thing knowledge"—knowledge that is encapsulated in devices or otherwise materially instantiated (2004). Baird uses the example of scientific instrumentation, which contains within itself skilled scientific knowledge. This second view of technological knowledge focuses on the actual objects created and used, rather than the clever humans who make them work. Though the scientists and technologists who build scientific apparatuses play a large role in making their knowledge visible and tangible by means of the instruments they build, the knowledge itself can be demonstrated by or used in intelligent ways without verbal expression.

In these accounts, know-how is not embodied only in people but also in the material products of science; materials themselves can provide explanation. Baird classifies thing knowledge into three different subtypes: model knowledge, working knowledge, and instrumentally encapsulated knowledge. Although Baird's notion of working knowledge is deeply rooted in the context of scientific ideas and practices, it may be relevant to some of the animal cases I examine: working knowledge is know-how that is demonstrated or instantiated by the construction of a device that can be used by people or creatures without the advanced knowledge of its creators. Baird's justification for thing knowledge, and his five ideals for knowledge (detachment, efficacy, longevity, connection, and objectivity; 2002), offer a framework for evaluating the material creations of non-human animals. In some animal cases, the knowledge embodied in an animal artefact is the product of evolutionary processes; as we will see, the beaver's dam may demonstrate more knowledge about the world than its tree-felling behavior, which is an instinctive response to the stimulating sound of rushing water.

BACKGROUND: BIOLOGICAL, ENVIRONMENTAL, AND ANTHROPOLOGICAL LITERATURE

The recent spike in studies of animal behavior has opened up philosophical discussions about animal minds, and many compelling case studies complicate our human-centered approach to technology. For the purposes of this book project, I choose to focus on only a handful of the existing cases—largely field studies. Though laboratory studies tell us much about the capabilities of certain animals, they do not tell us much about how animals actually behave in the absence of human direction. In this book, laboratory studies are used to bolster field studies of animal behavior, helping to build a case for inferences of purpose or intelligence in animal construction behaviors. Purpose and intelligence underlie many definitions of technological knowledge in the epistemology of technology literature; intentionality plays a role in every account of technology I've encountered. Though serendipity and exploration can lead to invention, the *intention* of using a thing remains fundamental to defining that thing as a tool. In my investigation of animal cases and how they might map to our epistemic accounts of technology, I've chosen case studies about five main groups of wild animals, and I include other contrasting cases and laboratory studies along the way.[4]

Perhaps intention is the definitional element that will allow us to decide whether or not animals have and use technological knowledge. Philosopher of language and biology Ruth Garrett Millikan (1984) has examined intentionality and the behavior of animals, particularly bee dances. Millikan argues that intentionality is relevant only in terms of a particular organism's evolutionary history. According to Millikan's argument, if a random collection of molecules were to serendipitously agglomerate and create an exact duplicate of me, I would not be able to refer to the duplicate's "eye" or "brain" or "intentions," because all of those terms refer to organs or mental states produced by my particular evolutionary history, which this random copy of me does not share (Millikan 1984, 93). Millikan's analysis contradicts certain pragmatic analyses in the philosophy of technology, which define things by how they function. Normally, we call something an eye when it does the sorts of things an eye does, and we refer to technologies with regard to their functions; for example, a discarded old clothes iron that is used as a doorstop *is* a doorstop. While Millikan's arguments provide some grounds for claims about intentionality in animal cases, her focus on language is problematic, because we cannot ask animals what they are thinking. In this study, then, tool use becomes an important marker of intentionality; it offers a way around the language barrier. Millikan's analysis provides an entry point for the animal cases I examine. Some of Millikan's conclusions are limited by her very narrow use of animal case studies; however, when her work is examined alongside the wealth of new observations of animal tool use and behavior, many of her arguments now map better to animal cases than she might have imagined possible.

If intention isn't quite the answer, perhaps this study should examine how material objects and environments encode technological knowledge; this would allow for animals' technology use and would obviate the need for intention. Environmental philosopher Mark Rowlands (2005 and 2013) has developed an environmental epistemology that, like Millikan's, recognizes that evolution and environment shape the ways that organisms navigate the world. Rowlands cites several examples from both human and animal cases indicating that the world itself is used like a memory store (2005, section 7). He explains that "the environment contains structures that carry information. This information is relevant to the accomplishing of a cognitive task, and it can be appropriated by organisms that are capable of acting upon the structures in the right sort of way" (ibid.). For both humans and non-humans, environments structure thinking. Some animals may respond to their environment's informational structures automatically, rather than intentionally; other cases, such as animals' construction of material artifacts, are more complex. When creatures use tools, the relationship between the creature and the environment changes; the environmental cues to which the creature responds are influenced in turn by the creature's shaping of that environment. Rowlands, like Millikan, brings our brains and behavior back to environmental context. Rowlands's idea that the world contains information about itself—that the world is its own map—is being applied in AI guidance systems for self-driving cars, and it seems to echo Baird's ideas about scientific knowledge being encapsulated in scientific instruments designed for particular purposes. Indeed, one might say that evolution and environments have produced organisms and structures that bear information in parallel ways.

Or perhaps animals can be said to have technical knowledge if they use technology for problem-solving. According to historian of technology Rachel Laudan (1984), technological progress can be tracked by examining the cognitive changes it produces in its practitioners, foremost among them the rise of problem-solving, which "constitutes the major cognitive activity of the technological practitioner" (1984, 84). Laudan offers a taxonomy of human technological activities. However, Laudan's framing of technology in terms of the problem-solving capacities it evokes presents significant problems for some of our animal cases. Spiders spinning webs, for example, cannot be said to be actively engaged in problem-solving (or can they?), but laboratory studies of chimpanzees and of crows (Bluff et al. 2007) suggest that these animals do problem-solve.

Might problem-solving, then, serve as some sort of cognitive dividing line, helping us to sort out whether these animal cases represent technology use? Inferences about problem-solving can be difficult to judge, however; even humans solve problems by luck rather than intelligence at times.

Another possible characteristic marking out technological knowledge is innovation. Anthropologists Schick and Toth (1993), who primarily study apes, survey tool use among animals, looking at innovation as a possible dividing line between technological knowledge and instinct. According to Schick and Toth (1993), sea otters that use tools to open mollusks are demonstrating innovative behavior, but birds that drop mollusks onto rocks to open them are relying on instinct—not engaging in the sort of flexible, innovative tool use observed in the great apes (humans included). These researchers see tool use as exclusive to mammals, which have more highly developed cultures, and they altogether dismiss birds' and insects' capacity for innovative tool use (1993, 55). When Schick and Toth published this study, there were fewer case studies of animal tool use—in fact, they argued that tool use among non-primates was rare (1993, 54)—but there are many more documented instances of animal tool use today. Their conclusions about the differences between tool use in different species should be updated to reflect the many, many cases of sophisticated tool use that have been observed in crows.

This book offers a new integration of material from multiple fields; I evaluate the new, exciting studies of animal tool use in terms of technological knowledge, and to do this, I use the definitional characteristics—intention, technological products that encode knowledge, problem-solving, innovation—put forward by previous researchers in various fields. This requires me to address questions about animal minds—whether animals set intentions and how intentionality evolved, whether animals are able to innovate, whether they can problem solve, how they learn—as well as questions about what constitutes *technology* and what constitutes *knowledge*. This book is predicated on the idea that technological knowledge really is a type of knowledge—an idea that has been under-recognized in the history of philosophy, which historically has valued propositional theoretical knowledge.

BACKGROUND: ANIMAL CASES

The first set of animal studies I focus on examine how chimpanzees use tools: in the wild, in laboratory situations, and in zoos. Chimpanzees have been observed using many types of tools; for example, Crickette Sanz has found that chimps, when raiding army ant nests, use a set of tools to avoid getting stung by the ants (Sanz et al. 2007, Sanz et al. 2009). Sanz and her team recovered more than 1,060 chimpanzee tools and recorded 25 videos of this behavior. The researchers themselves call these tool kits "technology" (similarly, as we will see in the corvids chapter, researchers into New Caledonian crows' tool use call the birds' creation and use of sticks for particular purposes "technology"). Though calling something a technology does not make it so, Sanz's videos of chimpanzee tool use are extremely compelling; the social factors involved in the chimps' tool use, and their sophistication and purposeful employment of these tools, pose a quandary for those who rule out non-humans' ability to use technology. Further research into chimpanzee populations, presented in Kathy Schick and Nicholas Toth's 1993 *Making Silent Stones Speak*, suggests that chimps in captivity can create new ways of carrying out a task without being shown how by researchers; this type of ingenuity is another important element used to define technology, and, by this measure, it seems that chimps (and, as we will see in a later chapter, dolphins) are capable of using and making technology.

While the chimp cases are striking, their use of technology is perhaps less controversial than that of other types of animals that the book examines; we tend to casually think of apes as being more like us than other creatures, so their uses of technology do not seem especially surprising. My second group of animal case studies examines bottlenose dolphin families' use of a peculiar hunting technique called "sponging," observed by Krützen et al. (2005). Eight dolphins were observed breaking off the top of sea sponges and using the pieces over their noses to scrounge up small fish off the ocean floor to eat. This seemed to be a shared hunting technique, with know-how passed among females in a family group. Other dolphin studies corroborate dolphins' capacity for this type of learned, communicated behavior (Janik 2000, Pryor and Norris 1991, Tyack 2000). After examining the Krützen study in some detail, I look at other dolphin foraging techniques and communication in both dolphins and whales to give further evidence for their know-how.

Even the dolphin cases may not seem revolutionary; we have seen dolphins' cleverness in fantastic water shows and TV programs. The third set of animal cases I examine, the corvids—primarily New Caledonian crows, who make sophisticated tools to catch prey—may be less easy for most readers to accept as being examples of technological knowledge. However, these cases are the most compelling of all, and they make up the heart of this book's argument.

Drawing from the work of Gavin Hunt and his team, which describes New Caledonian crows creating and using many types of tools for very specific purposes, I demonstrate how intentionality maps to animal cases. According to Hunt (1996), the tools that New Caledonian crows manufacture and use meet three criteria for tool use that were once originally thought to apply only to hominid tool use: standardization, the clear making of a form, and hook use (Hunt 1996). The crows keep track of their tools and use them repeatedly, reusing particular tools for particular purposes. Other bird cases reveal interesting things about the importance of environment in the development of tool use.

Finally, I examine two particular animal productions, spider webs and beaver dams, both complex material animal creations that appear to be instantiations of working knowledge. By drawing from these cases, I show that a complete epistemology of technology requires a more nuanced approach to the material products of non-human animals than is currently available.[5]

STRUCTURE

In the next two chapters of this book, I lay the groundwork for interpreting the animal cases I examine in the final chapters. In chapter 2, I define my terms (tool, technology, artifact, knowledge, etc.) and argue that there is no *a priori* reason to exclude all non-humans from our definitions of technology. In chapter 3, I take a closer look at technological knowledge in order to help situate the animal cases I examine in the remaining chapters. Chapter 4 considers the social learning of chimps and their tool-related behaviors. Chapter 5 examines dolphin sponging, culture, and learned know-how. Chapter 6 deals with the variety of material products that New Caledonian crows and other birds produce, and the apparent problem-solving that their tool-making encodes. Chapter 7 examines beaver dams, spider webs, and other animal constructions from the perspective of thing knowledge, looking at the role that environmental niche, design, and evolutionary processes play in non-humans' tool use. In chapter 8, I provide a theoretical framework into which all the previously described instances of animal construction and technique can be placed. This theoretical framework posits two axes on which construction or technological behaviors can be mapped: know-how (or learning requirements) and encapsulated knowledge. The map produced in chapter 8 shows that animal artefacts and technological behavior fit within the same spectrum of intentional technological behavior as human technological activities do. We don't consider all human technology and tool use equal; some activities are more sophisticated (higher levels of built-up-ness), some evidence more intensive learning processes, and some construct more original knowledge. The same should be true of tool-related animal behaviors. Providing for a spectrum of possibilities should help us at least consider animal and human objects and behaviors in the same terms.

With this book, I hope to induce philosophers of technology to consider animal cases and to induce researchers in animal studies to think about animal tool use with the apparatus provided by philosophy of technology. Readers may thus choose to read this book in different ways. Readers interested strictly in technological knowledge should read chapters 3, 7, and 8. Readers who wish to know more about animal tool-use behavior and do not care about technological knowledge should read chapters 2, 4, 5, 6, and 7. Readers well-versed in animal cases who want to engage with philosophical frameworks should turn to chapters 2, 3, and 7.

NOTES

1. View the video here: https://www.youtube.com/watch?v=Snl4U8HPdtQ. Other viral videos include crows sledding (https://www.youtube.com/watch?v=gjgpenWavO8) and orangutan reacting to a magic trick (https://www.youtube.com/watch?v=FIxYCDbRGJc).

2. Or "animal awareness." The problem of whether or not animals have minds is as vexing as the problem of whether human beings have minds. Strict materialists will say that there is no such thing as a mind, that we have only brains. I do not wish to engage this discussion. I do think that we can talk about animal minds in the same ways that human minds are discussed: in terms of intentions, thoughts, cognition, and/or awareness. The terminology of "minds" may be controversial in some circles, but it handily covers a wide range of mental content and representation. I use the term to refer broadly to the thought-content of animal brains (or whatever does the thinking in both humans and animals).

3. In my second chapter, I draw distinctions between knowledge, artifact, tool, and technology, and these definitions help to clarify this project. Though there is more to consider about animal artifacts and how they fit into philosophy of technology, I have chosen to concentrate on technological knowledge as the lens through which I examine animal tool-making and tool use. Technological knowledge, which has been treated fairly thoroughly in the philosophical literature, provides a relatively straightforward point of entry that, for example, metaphysics or the ethics of technology would not.

4. In "What Dolphins Want: Animal Intentionality and Tool-Use" (Shew 2008), I analyze one way of making sense of animals' material products.

5. I begin describing this project in "Spider Webs, Beaver Dams, and the Sticky Wicket" (Shew 2007): it might be better to imagine tool use and technology as existing on a spectrum rather than as either/or categories. I return to this idea in the concluding chapters of this book.

TWO

Humans Thinking About Other Animals

A series of American commercials for the GEICO auto insurance company featured the GEICO caveman, a trademarked fictional character who was also part of a very short-lived ABC TV series in 2007. These commercials, which feature the tag line "So easy even a caveman can do it," show the caveman encountering prejudice and microaggressions. The GEICO caveman resents the prejudice and the stereotypes about him that he sees everywhere he goes. Every time the caveman sees or hears the GEICO slogan, his day is ruined. In one spot, the caveman is holding a boom mic on a film set where a GEICO commercial is being shot; when the spokesperson says the GEICO slogan, the caveman gets angry, casting down the boom and walking off set. In another spot, he sees the slogan on a sign and sighs with exasperation. When he is interviewed on the news about the slogan, the anchor asks whether something can be offensive if it's true and tells him that his species is less evolved.[1] These fictional ad spots comment on more than just car insurance; they also nod toward the ways in which humans view themselves as superior, considering other species, including proto-humans, less advanced.

THE HUMAN CLAUSE

My goal in this chapter is to disambiguate the relevant terminology—artifacts, tools, technology, and knowledge—in order to set up my argument that some non-human animal tool-related behavior should be seen as existing on a spectrum with technology. There is no *a priori* reason to bar non-human animals from the category of creators and users of technology. However, many accounts of technology have had what I term "the human clause": the idea that human beings are the only creatures that can have or do use technology.

The concept of *homo faber*, (hu)man the maker, sits at the center of many definitions of technology. Technology, taken through this lens, exists as a domain exclusive to humans. According to this view, other animals might be smart (or not), but no other species can harness the power of physical manipulation, creativity, and design that human beings can—the power that is embodied in our technologies. We humans are taken as pinnacle in the animal kingdom because of our technological power and its adaptability (to environments, to problems, to entertainment). This human clause is apparent in many texts theorizing technology. Founders of the history of technology Melvin Kranzberg and Carroll Pursell explain that:

> Modern physiology, psychology, evolutionary biology, and anthropology all combine to demonstrate that *Homo sapiens* cannot be distinguished from *Homo faber*, Man the Maker. We now realize that man could not have become a thinker had he not at the same time been a maker. Man made tools; but tools made man as well. (Kranzberg and Pursell 1967, 8)

In the same foundational volume, another historian writes, "Technology is as old as man himself. Man was evidently a 'tool making primate' from the day when the first human-like creatures roamed the earth" (Forbes 1967, 11). R.E. McGinn writes in the first issue of *Research in Philosophy and Technology*, the journal of the then-newly-founded Society for Philosophy and Technology, that technology be treated "as a form of human activity, others of which include science, art, religion, and sport"[2] (McGinn 1978, 180). Another philosopher and theorizer about the nature of technology, Frederick Ferré writes that "technology, like human creature itself, can be understood as natural in both these [aforementioned] senses, i.e., rising naturally out of the character of *Homo sapiens*, as evolved within nature" (Ferré 1995, 29). More popular tech writer John Leinhard, writing on ingenuity and technology, explains the inevitability of human technologies in reflecting, like Kranzberg and Pursell, on the relationship between our bodies and minds; he explains that "Technology has driven our brains. Our expanded physical capabilities made technology—extended toolmaking—inevitable. Technology has, in turn, expanded our minds and fed itself" (Leinhard 2000, 4).

Following up on the idea of ingenuity in human work, Joseph C. Pitt proposes "*technology is humanity at work*" (Pitt 2000, 11, his emphasis). Noted historian Thomas Hughes also sees "technology as a creative process involving human ingenuity," writing more specifically about what we follow as people reflecting on technology:

> I see technology as craftsmen, mechanics, inventors, engineers, designers, and scientists using tools, machines, and knowledge to create and

> control a human-built world consisting of artifacts and systems associated mostly with the traditional fields of civil, mechanical, electrical, mining, materials, and chemical engineering. (Hughes 2004, 3–4)

More recently, Keekok Lee, writing on defining technology, provides the following description:

> Technology—in the general sense of the manipulation of nature to suit human purposes—is not peculiar to modernity. Technology has always existed since the first adze made by our Stone Age ancestors.[3] (Lee 2009, 19)

It seems that historians and philosophers of technology have limited their definitions of technology to human beings for as long as their academic disciplines have existed. Although most of these researchers accept that chimpanzees use rudimentary tools, technologies are almost always discussed as human projects.

One might reply: *why shouldn't they be described that way?* Humans make the definitions; we defined technology, and we should be able to insert the human clause wherever we like. I see no problem with this sort of response. At the end of this book, you may end up wanting to revert back to older accounts of technology that depend on the human clause. For the most part, perhaps, we can proceed in our philosophical reflections on the technological societies in which we live as we have been; these too are good projects. The philosophy of engineering may continue to grow as a discipline under these conditions, but I suspect that the human clause actually inhibits us from creating a broader account of the sorts of things that are constructed and manufactured in the world. To assert the human clause without first looking at the sorts of things some animals make and do and plan and implement is to do a disservice to a large segment of living things on Earth.

Our accounts of technology will remain impoverished if we neglect to look at what other species do. We should at least consider the new and exciting findings from animal research. Too often we consider humans separate from the larger world—we think of ourselves as special, and we justify this human specialness by reference to our unique facility with technology and language.

Before limiting technology to humans only, we ought to consider what it is that animals do and make and use. When the human clause is omitted, some definitions and accounts of technology actually map quite well onto what some animals do. Disrupting our inherited accounts of technology and their ties to humanity may be problematic for those who think that technology is what makes us human; however, I ask you to suspend the human clause for a moment so that we might at least take stock of what other animals actually do with technology.

In animal studies, one is instructed to be constantly vigilant against anthropomorphizing animal behaviors. Jane Goodall had to push back against editors who wanted to replace her use of "he" and "she" with "it" for her first publication (Haraway 1989, 409), but sometimes the treatment of animals as things is still apparent in our language. Nature writer Midge Raymond explains:

> New York Times editor Philip B. Corbett wrote in a February 2, 2016, article that the Times uses "person" pronouns "only for animals who have been given a name, or in cases where the sex of the animal is specified. Otherwise, we stick with 'it' and 'that' or 'which.'" . . . But for those of us who do write with a point of view, names and pronouns are important. . . . The scientists can continue to resist anthropomorphism—but this won't save the animals, or make the rest of the world pay attention. (2016)

Naming conventions for animals in scientific research have changed in part due to Jane Goodall's impact (Raymond 2016), and it is now much more common for researchers to have names for animals, though this varies significantly by species, with mice used for research much less likely to be named in scientific publications than in studies of apes and now even birds. Raymond worries that this puts them too far distant from our consideration and further "others" animals from humans. Humanizing animals matters to our moral consideration of them. When we lump particular species together, it is easier to wipe them from serious consideration. Taking individuals as important matters to both our moral thinking and, I would argue, to our thinking about epistemology. Of course, zoo animals, companion animals, and working animals are also given names by their humans, but that we think anthromorphically about these animals presents fewer problems because no one expects a person to be objective about their own house cats.[4]

In our fear of anthropomorphization and desire for a sparkle of objectivity, we can move too far in the other direction, viewing human beings as removed from the larger animal kingdom. Humans are animals too. I focus in this volume on technology, tool use, constructions, and techniques with the hope that focusing on other species' material and technical *products* will help us stay away from dangerous inferences of mind. By putting human technology in concert with other tool-related behaviors—the use of objects in one's environment to get to some end, the making of a form, the repeated manufacture or use of an object, the cultural transmission of tools and behaviors, etc.—we can be careful not to anthropomorphize or swing too far in the other direction. A hierarchical assumption about the relation of humans to the rest of the world (animal and environment) underlies much of the discourse surrounding the ethics and epistemology of technology. By considering animal cases, I hope to instead allow an entryway into considering material production and knowledge on their own terms, without implication about the minds that created them.

MAPPING OUT TECHNOLOGICAL BEHAVIORS[5]

A solid definition of tools can be found in *Making Silent Stones Speak* (1993), by anthropologists Kathy D. Schick and Nicholas Toth: tools, in their simplest form, are external objects used to achieve some goal (48). Tools may be modified or unmodified by their users, who, for Schick and Toth, will almost always be humans or protohumans. Artifacts, for these anthropologists, are a type of tool that has been modified in some way by human beings. However, they think even this definition is problematic, pointing to cases of chimps modifying tools such that they might count as artifacts. Technology is more than mere tool use: it "refers to the system of rules and procedures prescribing how tools are made and used. In a broader sense, this term can be used for the systems of tool-related behavior of non-human species as well" (Schick and Toth 1993, 49).

Schick and Toth note that "technological behaviors" are often more instinctual for non-mammals (51). This phrase, "technological behaviors," is one that I want to lift from Schick and Toth. Tool use and manufacture (the production of artifacts) exists in a continuum with more involved technologies. As you'll see in subsequent chapters, the hallmarks of technology—intellect, intention, design, use of form, and so on—can be attributed to at least some non-humans.

In *Philosophy and Technology* (1972), one of the earliest and most comprehensive volumes in philosophy of technology, Carl Mitcham and Robert Mackey give three ways that "technology" can be defined: epistemologically, anthropologically, and sociologically. In the following paragraphs, I will describe accounts of technology using these helpful categories.

In epistemological definitions of technology, knowledge of how to do or make things is central. Anthropological approaches center humanity in their definitions: for example, historians Melvin Kranzberg and Carroll Pursell, quoted earlier in this chapter, use an anthropological definition when they say that technology is fundamental to the sort of beings humans are. Schick and Toth, whom I have previously quoted, are anthropologists talking about technology, and so their definitions come from their field. Sociological approaches are "characteristic of thought and action in modern society" (Mitcham and Mackey 1972, 2), and these also tend to include the human clause. For example, philosopher Joseph C. Pitt, in his *Thinking About Technology* (2000), defines technology as "humanity at work"; here Pitt develops a sociological account of technology that involves the use of feedback loops to bring about improvements over time, and sociological definitions also ground Pitt's particular interest in philosophical accounts of engineering. Another example of a sociological approach is that of Andrew Feenberg, a prominent philosopher who examines the importance of sociological factors in modern technologies (Feenberg 1999).

Philosophers, however, often mix these three approaches in formulating their definitions of technology. Don Ihde, in *Philosophy of Technology* (1993), purports to define technology both "anthropologically and philosophically" (51). Technology, for Ihde, involves humans "relating to their environment" (51), and he focuses his anthropological-sociological definition on humans' non-neutral power to transform their environments. Frederick Ferré, in a book also entitled *Philosophy of Technology* (1995), provides an anthropological-epistemological account. Though Ferré does not restrict his definition to humans from the outset, his definition, "the practical implementation of intelligence" (26), can only be made relevant to non-humans by looking at anthropological or biological studies to see whether an artifact was created by the use of intelligence or by "mere" instinct. This is a more nuanced anthropological approach, and one that is an opening wedge for considering the sorts of technological behaviors of non-human animals that I will discuss in subsequent chapters.

None of these approaches is without merit, but each identifies something slightly different as an important feature. To create a sort of "map" that will help us distinguish between simple tool use, tool making, tool manufacture, and the use of technology, we can use elements of all three types of definitions. The use of instinct, intellect, and reason as ways to gradate technological behavior could be considered epistemological-anthropological. Sociological factors in technological behaviors include the social learning and culture that go into the transmission of technological behavior; epistemological factors include the use of know-how in the using and designing of tools; and anthropological factors include the evolutionary processes and intentionality that go into technological projects. I discuss these possible factors in the following chapters, paying special attention to the ideas of embodied knowledge and know-how. As a way to start piecing together this map, I will spend most of the next chapter setting up technological knowledge.

My proposal in this book is fairly simple: let's examine the animal cases to help us put together this map, this philosophical tool—this spectrum that (I hope) will enable cross-talk among philosophies of technology and will help situate human technology and animal construction in relation to each other. Rather than position humanity as being superior to other animals because of our use of tools and technology (the outcome of so many reflections on humans, animals, and tools!), I hope to show that the sorts of technological skills humans have are related to technological behaviors found elsewhere in nature, some of which may actually be much better adapted to their niches.

TOOL USE AND ANTHROPOMORPHISM

Benjamin Beck, in his often-referenced 1980 book *Animal Tool Behavior*, provides the most referenced definition of a tool: A tool is some external object that is modified, carried, or manipulated to effect some action. This definition discounts spider webs as tools because the spider produces silk from its body, not an external object; it also distinguishes between tools and constructions such as nests and homes, because constructions cannot be detached from where they are built. This definition, which is still widely used, does have some detractors, which I will discuss in chapters 7 and 8.

One objection that I need to address before continuing is the fear of anthropomorphism. Our definitions of technology have been tightly associated with the human. Our academic work reiterates the human clause; in popular news, commercials (like the GEICO caveman ads), and cartoons, we have been inundated with imagery that links technology with humanity. However, serious anthropological work with chimps and newer studies on crows and dolphins demonstrate that other animals use tools, make tools, convey techniques, and use or construct artifacts. Are these animals creating technology? Well, that depends on how we define technology, but they are certainly doing something complex that ought to be at least regarded as interesting and recognized by philosophers of technology in their discussions of what technology is. To neglect constructive behaviors simply because humans are not doing the construction is to neglect a rather large segment of made things and processes of making. At the very least, the nature of technology is not merely tied to the nature of humanity, but to *humanity in our animality*. The nature of technology speaks not to the nature of humanity, but to the fact that humanity is coextensive with animality. Philosophers of technology who focus solely on humans and human projects reinforce the view that the human exists outside of nature, even as they argue against the natural/artificial distinction.

Other animals are unlikely to think about themselves as users of technology or tools (at least in any way that I might understand, though dolphins have been documented to understand abstract categories, see chapter 5). They are using a hunting strategy or shaping a stick with the idea of getting prey, whether or not they conceptualize it thus. We cannot know what it is like to be another animal, so positing animal thoughts may not be possible; by the same token, however, we can never truly know the mind of another human being, and every day I posit the intentions and feelings of other human beings. I do this by looking at *what other people do and make*. While we cannot assume mental content without very careful study, we *can* look at what creatures are making and doing and passing on to one another. By following the objects themselves as they are produced in the field, and by combining these observations with laboratory studies about the abilities of animal groups, we can come up with decent accounts of what is made and done in order to situate these behaviors in relation to technology.

Over the past fifteen years, there has been an explosion of research into cognition and emotion in the field of animal studies. With this research has come skepticism—the concern that researchers have been inappropriately ascribing human categories to non-human animals. But the concern about anthropomorphism can drive us to swing too far in the other direction, ignoring the intellectual, sensory, and emotional landmarks that shape how we treat others (including animals). Much of the flurry and fuss over recent animal studies has been about animals' emotional (and—relatedly?—moral) standing. This is not my project. I address the problem of anthropomorphism, but in the end I will have little to say about how we should treat other beings; this project focuses instead on how we should think about the sorts of things humans and animals do and make, and about how these doings and makings can be considered together in a single epistemological account.

Many thinkers and writers have addressed the problems of anthropomorphism and the attribution of mind that drives the phenomenon. In a non-technical, personal account of his life with his dog, nature writer Ted Kerasote explains that there are inaccuracies in both ascribing and in not ascribing human characteristics to animals:

> Anthropomorphism is often maligned for ascribing human characteristics to animals who can't possibly know what we know. And there is some truth to this. I doubt Merle thought of the Big Bang when he gazed at the starry heaven. But the reverse—not ascribing volition to creatures who repeatedly display it—is also inaccurate. It leads to what poor translation always does: misunderstanding between cultures. (2007, 112)

Skeptics would counter, perhaps, that even the use of "culture" here is an improper ascription of human characteristics to animals. However, we routinely assign human characteristics—the categories with which we think about our own personal experience of the world, our own mental traits, and our own behaviors—to the sorts of things other human people do. David Hume, in *A Treatise of Human Nature* (1888), makes the case that the difference between non-human animals and humans is a matter of degree, and not of kind:

> No truth appears to me more evident than that beasts are endow'd with thought and reason as well as men. . . . We are conscious, that we ourselves, in adapting means to ends, are guided by reason and design, and that 'tis not ignorantly nor casually we perform those actions, which tend to self-preservation, to the obtaining of pleasure, and avoiding pain. When therefore we see other creatures, in millions of instances, perform like actions, and direct them to like ends, all our

> principles of reason and probability carry us with an invincible force to believe the existence of a like cause. (176–177)

Hume continues, explaining that there is a great resemblance between the external actions of animals and our own actions, and this resemblance should cause us to see animals and humans as having mental operations in common (177). In this way, Hume argues that reason, custom, and instinct operate in both men and animals (179).

Mark Bekoff and Jessica Pierce (2009), in a discussion of ethics, make the claim that it is narrow-minded to discount the possibility that animals think and feel:

> New information that's accumulating daily is blasting away perceived boundaries between human and animals and is forcing a revision of outdated and narrowminded stereotypes about what animals can and cannot think, do, and feel. We've been too stingy, too focused on ourselves, but now scientific research is forcing us to broaden our horizons concerning the cognitive and emotional capacities of animals. (Bekoff and Pierce 2009, x)

I'll be presenting some of this new information from animal studies that brings into question whether categories like culture, intelligence, mental time travel (in psychology), and planning are exclusive to human beings. Perhaps animals are best imagined as Ted Kerasote views them, as speakers of foreign languages: translation is difficult and takes a concerted effort, and deciphering meaning is no easy process, but meaning is there (2007, 10–12). Anthropomorphism—the ascribing of human categories to non-humans—may sometimes be too hasty or misguided or in error, but if care and patience are used, anthropomorphism can in some cases help us figure out what another creature is doing.

In December 2005, the *San Francisco Gate* first reported on the rescue of a humpback whale (Fimrite 2005) and the tale was sent around as an email forward and re-reported in publications across the country. A female humpback whale who had gotten herself tangled in crab lines near the Farallon Islands (about eighteen miles from San Francisco) was discovered early in the morning by a crab fisherman, who called the Marine Mammal Center. The center sent out a group of divers to evaluate the situation and try to rescue the whale. The divers were not sure they would be able to rescue the whale, who was tangled in the lines and heavy crab traps, and the rescue was dangerous "because the mere flip of a humpback's massive tail can kill a man." Around the whale hung twelve crab traps weighing ninety pounds apiece, and the whale was struggling to keep is blow-hole out of the water. The divers spent hours cutting the ropes—while the whale held still for the divers. *SF Gate* reports that:

> When the whale realized it was free, it began swimming in circles . . . it swam to each diver, nuzzled him and then swam to the next one. . . .

> Whale experts say it's nice to think the whale was thanking its rescuers, but nobody really knows what was on its mind.

One of the diving team organizers explained that, "You hate to anthropomorphize too much, but the whale was doing little dives and the guys were rubbing shoulders with it . . . I don't know for sure what it's thinking, but it's something I will always remember." The worry about anthropomorphism keeps us from making large generalizations from an incident like this; however, while we may not wish to conclude that this whale was showing gratitude, it might be useful to describe the whale's behavior using human terminology: patience or carefulness (in holding still) and excitement (as witnessed in his little dives and nuzzles against the divers). We make these inferences quite often about animals with whom we are familiar, on the motivations of cats and dogs, for instance.

Mike Hansell, in *Built by Animals* (2007), explains why anthropomorphism became a huge concern among animal studies researchers and why questions of whether animals and humans shared moods or behaviors went unaddressed for some time:

> I was warned as a student that anthropomorphism was a sin against science. In 1872 that brilliant and versatile scientist Charles Darwin published his book *The Expression of Emotions in Man and Animals*. This includes, among many skilful etchings by a certain T.W. Wood, the face of a chimpanzee bearing the caption "Chimpanzee disappointed and sulky." Soon after this it became impossible for a scientist in the field of animal behavior to say such a thing and still retain their scientific credibility, and remained so for the best part of a hundred years. (2007, 8–9)

The reason, according to Hansell, for this shift against anthropomorphism had to do with the frustration of scientists during the early twentieth century where "fruitless debates on the relationship between mind and body generated by the approach of so-called psychological introspection (what you might call 'thinking about thinking')" (8–9). Those researchers in the early days of the discipline of animal behavior hoped to establish animal behavior as a clear science with stress on experimental objectivity, and therefore rejected methods of introspection which seemed less scientific and objective. Hansell goes on to explain that it was not until 1976, when *The Question of Animal Awareness* by Donald Griffin was published, that the field of animal behavior research seemed mature enough to tackle questions about the nature of animal minds. The previous taboo against considering the thoughts and feelings of animals in behavioral research is now lifted, and with this shift, we've witnessed a flood of animal mind studies that would have once been ridiculed.

The message of this new perspective, says Hansell, is that questions of whether animals think and feel are now within the bounds of scientific inquiry (2007, 10). I draw from some of this recent animal studies material with the acknowledgment that it's possible to go overboard. However, if approached with care, anthropomorphism is not only justified; it is actually desirable if we indeed hope to make sense of the world and its inhabitants.

In this book, I apply concepts that were once used only in philosophy of technology to the material creations of non-human animals. There is justification for this in the chapters that follow. Briefly, discounting these animal material creations because we assume that technology is unique to humanity means that our accounts of material production, knowledge, and artifacts will remain isolated from a much larger discourse, one that could also take account of the information and production of non-humans. This book is aimed at bringing these discourses together.

The GEICO caveman demonstrates one important thing: the constant lack of respect for the achievements and intelligence of others who are not like us cuts too far. We need to take animals seriously as technologists to provide comprehensive accounts of what technology is and means. This book works to rectify some of the ways in which philosophy of technology has neglected the material creations of non-humans.

NOTES

1. A wonderful compilation of all these GEICO commercials can be found on YouTube: https://www.youtube.com/watch?v=4CF9dn4OJsw.

2. This is followed by the parenthetical remark: "It should be added that recent work of Jane Goodall and associates has shown that humans are not the sole practitioners of technology. Not only do chimpanzees use tools—e.g., compress leaves, inserting them in water-filled tree trunks as sponges—but they also engage in rudimentary tool manufacture: stripping leaves, inserting them into tree trunks to extract termites." Nonetheless, the author continues to analyze technology as a human activity, despite recognizing evidence that should expand his scope.

3. The author here thinks technology begins only when early human ancestors started making tools.

4. For a summary of animals in philosophy of technology, see Shew 2017 in the forthcoming *Spaces for the Future*.

5. Later I will be using this term almost interchangeably with construction behaviors; things like webs and nests are not considered tools, but animal constructions, since they are not easily detachable from their environments.

THREE

Technological Knowledge

This chapter provides an overview of the topic of technological knowledge in order to frame how animal studies can be used in philosophy of technology. Marc J. de Vries, in "The Nature of Technological Knowledge," lamented in 2003 that there were not yet many philosophical studies of the nature of technological knowledge, and his statement remains largely true. Most of what is now published and presented on technological knowledge is in the sub-domain of engineering knowledge. According to one acquaintance at a Society for Philosophy and Technology meeting (2009), technological knowledge used to be a hot topic, but has now fallen out of favor. Here, I use the concept of technological knowledge to frame my discussion of the types of things humans make and do and the types of things non-human animals make and do.

In my introduction, I described two broad types of accounts of technological knowledge: technological knowledge that resides within the maker, designer, or user of an artifact (often called "know-how") and technological knowledge that resides within an artifact ("thing knowledge," in the terminology of Davis Baird). For the most part, I will call this first type "technological knowledge" and the second type "thing knowledge," though I hope to recognize that thing knowledge is also a sort of technological knowledge. These two types, though quite obviously connected, are hard to describe in relation to each other because their relation is not linear. Things tend to "bite back," in the words of Edward Tenner (1997), and so the relation of thing knowledge to technological knowledge as it is normally defined is a messy one. However, the literature on technological knowledge of both types may help us to approach the topic of non-human animal tool use. Subsequent chapters will consider the animal cases more closely; in this chapter, I will describe several ways of thinking about technological knowledge and thing knowledge, and how these ways might fit with the animal cases provided in later chapters. All the current accounts of technological knowledge found in history and philosophy of technology revolve around humans, human technology, engineering, and science,[1] so I hope to provide in this chapter a less particularized reading of technological knowledge.

Discussions of bodies of knowledge often are detached in some way from the objects of knowledge; as Joseph C. Pitt argues, it is impossible to talk about *technology simpliciter*, and we must instead talk about individual technologies (2000). We should still, however, be able to talk about technological knowledge. After all, although sociologists and philosophers of science dispute whether we can ever talk about science in the singular (because science is made up of vastly different disciplines), these scientists in the same breath discuss scientific knowledge, its construction, and its use. Looking at the nature of technological knowledge will enable us to look at the sorts of things animals make and do within the same framework of investigation as things human beings make and do.[2]

DESIGN, BIOLOGICAL AND TECHNOLOGICAL

The most in-depth and referenced study of technological knowledge to this point is Walter Vincenti's *What Engineers Know and How They Know It* (1990). Though the book is now twenty-five years old, Vincenti's analysis of the use and construction of knowledge by aeronautical engineers involved with design remains the most thorough study of technological knowledge. Although his depth is made possible by his concentration on a specific set of case studies in one branch of engineering, his discussion of categories of technological knowledge and design is relevant to more general discussions of tool use. According to Vincenti, there are three goals of engineering: design, construction/production, and operation, with design taken as central. Design, according to Vincenti, is "multilevel and hierarchical"—there are different levels of design that interact with one another (1990, 7). He distinguishes between normal and radical design: normal design exists where there is knowledge of how a device works and what should be done by the engineer, and radical design exists in situations where there is no pre-existing knowledge about how to work or design a device (7–8).

Vincenti describes normal design as "evolutionary rather than revolutionary" (8), a sentiment that is echoed in many historical studies of technology, including Henry Petroski's *The Evolution of Useful Things* (1992) and George Basalla's *The Evolution of Technology* (1988). The use of the evolutionary metaphor to describe technological change is not trivial. Petroski's account of the role played by failure in design follows evolutionary theorists' descriptions of the role of extinction and of mutations that have no survival value (Petroski 1985, 2003, 2006). Like failed biological designs (mutations without survival value), engineering failures may tell us more about a project—and about the specifications and external pressures for that project—than successes (Petroski 1992). Like evolutionary biology, engineering knowledge "reflects the fact that design does not take place for its own sake and in isolation" (Vincenti 1990, 11). What we think of as the biological design of, say, a spider's silk is actually the product of an evolutionary history that happened in the context of environmental pressures on some early arachnid. The pressures that drove the use of silk (and subsequent web-making, for some spider species) may have been climatological, geographical, a matter of prey and availability, etc. External pressures drive both biological and engineering design.

By the end of the book, Vincenti actually champions a model of the growth of engineering knowledge that comes from studies of biological adaptation. He argues that Donald Campbell's variation-selection model "is fundamental to all genuine increases in knowledge [including the knowledge generated in his aeronautical engineering case studies], from that embodied in genetic codes arrived at by biological adaptation to the theoretical structures of modern science" (1990, 241). Within the scope of Vincenti's work, the model works like this:

> Design itself constitutes a variation-selection process of knowledge generation. . . . In [a] more normal situation, the desired knowledge is how to arrange and proportion the particular device so as to accomplish its task given the constrains of the normal configuration. The design of most airplanes after the early 1910s was more or less of this kind. In this situation, the designer usually lays out a number of plausible variations on some basis and selects the final design by some sort of analysis or experimental test or combination of both. More often than not, the process takes place iteratively. (244–245)

This iterative process has been described elsewhere as a system of "feedback loops" with assessment feedback that allows for the improvement of technology over time (Pitt 2000, 13–15).

It seems a stretch at first to apply Vincenti's account of technological knowledge to the realm of animal behavior, because he is examining the high-technology realm of aeronautical engineering. However, his theory relies upon a variation-selection model to help explain how knowledge grows over time, and this offers an easy point of entry into biological cases. There is a strong, though rarely highlighted, connection between the literatures of engineering design and biological design. Though biological design, unlike engineering, does not center intentionality, the two fields share the concepts of "design" and "function"; these linguistic links provide an interesting way to approach technological knowledge, especially as it is described by historians and philosophers using evolution as a central metaphor in their accounts of technological change. Biological literature describes how spider webs are designed to catch bugs for prey without problems of intentionality creeping in. Biological function is how something does what it does, and we think of engineering function in almost the same terms: a function is what something does. A web functions to catch prey; the function of this net is to catch butterflies. A web and a net are functionally similar ideas, though their designs take radically different historical paths, one through biological and one through technological evolution. Someone catching butterflies with a net has a very different goal than a spider catching flies in a web, and the learning process and construction techniques for these two tools are radically different (the spider requires no training).

Beyond this emphasis on function, there is a second way to bring Vincenti's technological knowledge to bear on the non-human cases; in those cases where we see evidence of *intentional* design on the part of some animals, we might talk about the design process and the transformations of knowledge that take place in very much the same terms as Vincenti uses to describe human aeronautical engineers' processes and learning. It seems that intentional design of this type is impossible for a spider, but there are many animal cases that do align with technological accounts of intentional design.

KNOWING HOW

Vincenti, using a distinction first made by Gilbert Ryle (1946), sees knowledge as including both knowing-that and knowing-how (1990, 13), and sees both of these types of knowing as critical to engineering knowledge. Knowing-how is described in Edwin Layton's multiple discussions of the differences between scientific knowledge and technological knowledge (1971, 1974, 1987). Layton, like Vincenti and others, sees design, or the ability to design, as important in defining technological knowledge. Design is, for Layton, "an adaptation of mean to some preconceived end" that results in made artifacts (1974, 37). For Layton, the design process involves a mental conception, followed by a more detailed plan, then the drafting of plans (such as blueprints or a model), and finally the making of some object or thing (1974, 37). For Layton, knowledge of design—its systems and products—is know-how (1987, 604), and the process of forming an idea and making it into concrete reality—the essence of design—is the "highest level" of knowing-how (604).

To examine non-human cases within Layton's framework, we must see whether non-human animals are engaging in the design process. Figuring out whether a tool-making animal has an idea of some end or object in mind may be tricky, but we can watch for and identify behavior that indicates prior consideration or planning. Other types of evidence of design could include repeated making of the same object, the sharing and learning of some technique, or evidence for animal planning for some event. In fact, non-human cases of all of these behaviors are described in later chapters.

Layton writes that technological knowledge "can exist on all levels" (1987), from abstract thinking about technology to specific objects. Technology does not involve *only* mental components for Layton, as know-how involves physical memory and training that cannot be adequately described by theory. Technological knowledge is used in all types of tasks, from the most difficult to the least intellectual; a typing clerk has the appropriate know-how for his or her job, though it is not theoretical in nature (1987, 604). The techniques and skills of workers represent know-how. However, Layton gives several odd caveats against taking too far these comparisons between physical know-how and abstract know-how: "I do not think comparing Einstein's knowledge of science with a child's knowledge of how to ride a bicycle advances our understanding of anything" (1987, 605). Layton is attempting to make the point that technological knowledge is "tailored to serve the needs of design," and that simple physical know-how is less central to the design process than abstract or theoretical know-how; for Layton, comparing two seemingly incomparable things adds little to our understanding.

Unlike Layton, I do think it makes sense to compare seemingly incomparable types of technological knowledge—knowledges that are at a great distance from one another. Recognizing the cultural and biological niches in which tools and technologies are employed is important to making this type of comparison appropriately. If we are to compare the know-how of, say, a beaver constructing a dam to that of a human constructing a dam, we should expect the types of know-how displayed to differ significantly. The knowledge or skills employed are vastly different and come from different ecological and environmental histories altogether. In the case of the beaver, the skills employed stem from some inclination or instinct in response to running water—an instinct that human beings simply do not have. Although we may never create a full, nativized account of know-how in beaver-dam building because we simply cannot understand what it is like to be a beaver, we can compare the processes and procedures and knowledge transformations of the beaver's construction to those of other construction projects. Identifying the differences in the two sets of processes makes for a fairer comparison. In this book, I hope to walk the line between the literature of engineering design and that of biological design, and the bridge between the two is the know-how involved.

THING KNOWLEDGE

Design, says Layton, is "manifested" in made artifacts (1974, 37). Know-how is realized in the creation of devices—it is *in the devices themselves*. Therefore, another way to approach technological knowledge is to look at its instantiation *within* technological products. Davis Baird (2004), in his account of thing knowledge, describes how the material products of science embody or bear knowledge, which is very different from the way in which scientific theories bear knowledge. While Baird specifically focuses on how scientific instruments bear knowledge, he also includes "other material products of science and technology," such as "recombinant DNA enzymes, 'wonder' drugs and robots," as knowledge-bearing things (2004, 1). Baird does not have in mind beaver dams or spider webs or the tools crows make and use (in fact, he specifically excludes spider webs). He offers, however, three ways that scientific instruments and material products bear knowledge: model knowledge, working knowledge, and encapsulated knowledge. These three modes of knowledge-bearing are worth exploring in relation to non-human animals.

Model knowledge, as Baird defines it, seems particular to scientific inquiry. For Baird, material models explain or demonstrate a theory, often better than a theoretical model might. Baird refers specifically to several physical models that have functioned thus: John Smeaton's model water wheel, used to determine efficiencies for waterwheels; Watson and Crick's "ball and stick" DNA models used to figure out the structure of DNA; and eighteenth-century physical models of the solar system, called orreries (2004, Chapter 2). Science's material models demonstrate or elucidate some phenomena by making visible some part of the world. To create a scientific model of this type, one must be in search of scientific knowledge, and science is a practice seemingly limited to certain groups of humans.[3] There is only one case in the animal literature that I think might qualify as this type of knowledge (see chapter 4); because model knowledge is so deeply tied to scientific explanation, witnessing it in animal cases is unlikely.

Baird's second type of thing knowledge is *working knowledge,* which he says can be held by both people and things. A person has working knowledge when he or she knows how to do something, and a device can be said to have working knowledge when it works regularly (2004, 45). To justify his claim that reliably working devices encapsulate knowledge, Baird appeals to pragmatism and the idea of *knowledge as effective action*. According to Baird, making material knowledge involves "contriving, arranging, and refining materials" (45)—in other words, effective action through use of a tool, something that is certainly seen outside of the human species.

A third type of thing knowledge is *encapsulating knowledge*. This type of knowledge, which combines model knowledge with working knowledge, is found in the creation of measuring instruments, which effectively de-skill a process. This instrumental type of knowledge requires both a phenomenon and a model from which to make a measure.

Though Baird's categories of thing knowledge are specific to scientific instrumentation and to the goals of scientific work, his general discussion of thing knowledge is highly relevant to a wider variety of cases. In a paper on the topic of thing knowledge (2002), he claims that "an artifact bears knowledge when it successfully accomplishes a function" (15). In order make sense of the classic definition of knowledge, "justified true belief," in the context of objects, Baird suggests that we replace truth with function in the material realm. When something functions properly, it is "true"—as expressions like "a true wheel" and saying something "runs true" seem to indicate (16).[4] Using a "thin" notion of function (by which he means one that is not loaded with intentionality), Baird gives us five ideals for knowledge that he thinks can describe scientific as well as thing knowledge (19):

detachment,
efficacy,
longevity,
connection, and
objectivity.

These five ideals help us to assess what counts as scientific truth; they also allow for arguments that some made objects should count as true, in Baird's pragmatic sense of trueness. Baird's ideals provide a way to talk about both non-material and material creations in terms of knowledge and truth.

For Baird, the notion of "detachability" is important for both scientific theories and scientific instruments: both should be detachable from their original contexts so that they can be applied elsewhere. Baird's discussion of the detachability of knowledge also suggests a link to information, although Baird does not make this distinction. By "information" I mean "a collection of facts or data," not "knowledge derived from experience" ("information" 2003). Facts or data can be stored without any active knower. Knowledge, however, requires a knower: a skilled worker or someone who understands a theory or its components (in the case of some large scientific theories, knowledge may be distributed). The "knowledge" that, according to Baird, makes it possible to "de-skill" large, complex machinery is in actuality "information," which is often processed by some technological non-human thing, usually a computerized device. I will further discuss this distinction between knowledge and information in chapter 8, but it relates to the ways in which something like knowledge about the world is evolutionarily encapsulated into the instinctual material productions of some non-human animals.

However, Baird does not see knowledge in animals' instinctual material productions; in fact, he specifically speaks against spider webs as encapsulating knowledge. He argues that objects made by humans can be linguistically and materially distinguished "from other natural products of life" (2004, 142). However, this assumes that humanity is somehow divorced from other life—a prospect I wish to deny vehemently.

While much of human creation is recognizable as human construction, there are cases that are less obvious: rocks that look like arrowheads, man-made ponds, fake stone that seems authentic, lab-created diamonds, etc. There are cases of human construction that are ambiguous, and there are natural phenomena that may be mistaken for products of human intervention. Baird advances a second argument against the prospect of spider webs as thing knowledge by appealing to his five ideals:

> Thing knowledge, existing in a more refined, constructed space, exhibits greater simplicity . . . than do the adaptive living creations of natural history. . . . [O]ur material creations, through our various acts of calibration, connecting them with one another and with what we say, have a greater depth of justification than do animal phenomena. Spiderwebs are well adapted to catch flies. But there is no connection established between this approach to catching spider food and other possible and actual approaches. We can and do connect direct-reading spectrometers with other spectrographs. (2004, 143)

A spider web, says Baird, cannot meet the ideal of connection—of establishing a relationship between the world and us—because spider webs are not related to other things. Baird's case against spider webs is intended to be a kind of straw man, an example that is easy to dismiss as obviously not having thing knowledge. Yet Baird's own five ideals of thing knowledge (detachment, efficacy, longevity, connection, and objectivity) speak against this dismissal of spider webs as technology. Detachment might be the hardest criterion for a spider web to meet: yet although spider webs are fragile, they can be detached or made elsewhere. These webs meet efficacy criteria, for they accomplish their ends. Webs meet longevity standards at least as well as some items that are unquestionably classed as technologies, for they can be depended on in the future. (I think here of an instrument that needs near constant fixing to stay in working order; I consider also the assumed disposability of many of the material creations of humans.) The connection criterion is also met, despite Baird's assertions: spider webs do establish a relationship to the world, a physical relationship that puts the spider at home, that anchors it in the world. Sometimes the world is its own map, and the connection here is clear. As for objectivity, the world's voice certainly has priority in the relationship between the web and the world, given the fragile nature of the web and the world's many possible threats to it.

While I do not intend to argue that a spider has technological knowledge, Baird's divorce of function from intention becomes highly problematic in the face of a wider range of animal artifacts. In the following chapters, I will provide cases that will help distinguish between intentionally created animal artifacts and those that, like spider webs, are largely the product of instinct, a sort of extension into the world of the spider's physical self—as Richard Dawkins explains in *The Extended Phenotype* (1982), "a temporary functional extension of her body, a huge extension of the effective catchment area of her predatory organs" (198). If we can distinguish between those things that are made by an animal out of some sort of phenotypic action, like Dawkins's spider web, and those things that are made with intention (with intention perhaps indicated by the sharing of some learned activity within a species—an activity that is not universal for that species in that environment), then we may have evidence of something beyond mere instinct at work. For Dawkins, animal artifacts such as ant hills and spider webs are simply extensions of that animal's body. But there are animal cases where tool use changes and evolves, where there are certain tools for certain tasks, and where techniques are socially transmitted.

PUTTING TECHNOLOGICAL KNOWLEDGE IN THE CONTEXT OF ANIMAL CASES

The chapters that follow are case studies of animal behaviors that are candidates for classification as technological knowledge, followed by (in chapter 8) a discussion of how we can fit these animal constructions into dialogues about technological knowledge. I construct from the case studies an account of technological knowledge that integrates knowing-how with thing knowledge. By integrating these two dimensions of technological knowledge, I hope to show how animal technological behaviors can be mapped alongside human engineering feats.

Many of the compelling animal cases come from the recent boom in animal studies research. According to Marc Bekoff and Jessica Pierce (2009), the past decade has been "the decade of the animal"; in their piece in the *Chronicle of Higher Education Review*, Bekoff and Pierce explain that "research on animal behavior has never been more vibrant and more revealing of the amazing cognitive, emotional, and moral capacities of a broad range of animals" (2009). In the humanities, this boom in animal studies has focused on morality and animals' under-appreciated emotional nature. This book offers a different focus: it integrates this new animal research with research in the philosophy of technology. I examine the animal cases in the context of technological knowledge because material artifacts and techniques are manifest—they do not involve attribution of mind or empathy, but allow us to examine behavior. While humans cannot know what it is like to be a bat (as in the classic philosophical thought experiment in Nagel 1974), we do know what it is like to use and make things, and we can recognize the constructions of other animals as being on the same continuum.

NOTES

1. Engineering studies is currently popular with thriving Forum on Philosophy of Engineering and Technology (fPET) workshops, the founding of the International Network of Engineering Studies in the late 2010s, and a continuing and intense focus on engineering in philosophy of technology.

2. Here I again ask that you suspend the human-claused use of "technological." I think the animal cases will speak for themselves in subsequent chapters, but, for now, I just hope to set up technological knowledge as a frame through which we can look at a variety of different cases—be those human or not.

3. Less problematically, we can talk about the "folk physics" of different species. When crows of the same type repeatedly drop nuts off a ledge to crack them open, it seems like something of an understanding of gravity and forces is being used—and animal researchers work on showing this too.

4. This, of course, harkens to the pragmatic work of William James on truth.

FOUR

Ape and Primate Cases

This chapter discusses previous research into the tool use of great apes and other primates. The literature on this topic is now well established, but the first scientific observations of non-human primate tool use were not made until 1960. Research into ape behavior documents material culture, social transmission, capacities for innovation, planning for the future, use of multiple tools in sequence, and indications of causal reasoning. These ape cases are therefore very well-suited for discussions of what counts as technological behavior. Apes, and their tool use, have received more scholarly attention than other groups of species. Many researchers assumed that non-human and non-primate animals did not use tools: why look for data about tools when you suspect none exists? But the fifty years of observation on great ape behaviors offer many cases of tool use. In this chapter, I highlight significant cases, particular animals, and important contrasts.

Tool use in apes is the least controversial and best documented of the animal cases. Many anthropological definitions of technology explicitly extend the human clause, converting it to a hominid clause (or at least noting the possibility); these definitions cite evidence of tool making by human ancestors and observations of tool use by orangutans and chimpanzees. In this chapter, I use findings about the great apes—gorillas, chimpanzees, and orangutans—to argue that the application of the human clause excludes too many tool users. Though many philosophers of technology mention chimpanzees' tool use, recent literature on the use of tools by chimps and other great apes has shown that their tool use is both more complex and more targeted than had previously been suspected. This new data makes maintaining a human clause in our definitions of technology much more difficult. In this chapter, I describe some of the pertinent research into the tool use of apes and conclude by examining how this research might be better incorporated into discussions of technology. This research gives us grounds to at least consider replacing the human clause with a primate clause, and it works to expand the domain of discussion about technological artifacts. In subsequent chapters, I will go on to argue that even a primate clause cannot be maintained in the face of other non-primate animal artifact production.

TOOL USE IN CONTEXT

In the 1960s, chimpanzees were observed using tools by Jane Goodall and her research group. Yet even as late as 1990, the data on primate tool use was limited: it was held that "in the wild only chimpanzees have been observed to make tools" (Boesch and Boesch 1990). Observing tool use in the wild requires human presence or observational equipment, and chimps are aware of this presence or equipment, often changing their behavior in response; habituating a group of apes to the presence of humans takes time. Thus, "in 45 months of initial observation, Goodall saw ant dipping only once, although it is common in Gombe" (Boesch and Boesch 1990).

Only recently have researchers been able to document the wide array of tool use by primates. Research published in 2005 documented gorillas' use of simple tools to perform tasks in the wild—using a stick to test water depths, using sticks and shrubs to support their posture, using a tree trunk as a bridge across swampy ground (Breuer et al. 2005). Orangutans have been documented using branches for foraging and using leaves to modify calls to mislead potential predators (Scheldeman 2009; Breuer et al. 2005). And chimpanzees, of course, have been documented using a wide array of tools: using stone tools to process food and crack nuts, hunting with the use of wooden spears, and more (Sanz, Call, and Boesch 2013).

These tool-related behaviors are learned through watching others and are passed within a community (Jane Goodall Institute 2007). The chimps of Gombe, who were studied by Jane Goodall, were found to use an array of objects—stems, twigs, sticks, leaves, rocks—in nine different ways. Chimpanzees in various areas have slightly different material cultures, making different tools for the same purposes. For example, chimpanzee studies by Crickette Sanz and her team in the Congo have revealed that some groups of chimps forage for army ants using "tool kits" that produce more sustainable ant-harvesting methods (Sanz et al. 2009); it is to these tool kits that I turn next.

CASES AND STUDIES

The Tool Kits of Chimps in the Congo

There are two types of chimpanzees: bonobos (*pan paniscus,* also called pygmy chimpanzees) and the common chimpanzee (*pan troglodytes*). Observational research in the wild on the common chimpanzee has yielded a wealth of new data on the type and purposes of tool use by chimps. Led by Crickette Sanz, an anthropologist from Washington University at Saint Louis, researchers have found that common chimpanzees employ a complex set of tools (not just one simple tool) to prey upon ants. Though chimps had previously been observed using dipping tools to extract ants from anthills, this new finding suggests that chimps can employ tools in concert to better extract their food (here "better" means "more sustainably"; Sanz et al. 2009). Sanz's research team has collected more than 1,060 tools used by chimpanzees for ant-dipping and has captured twenty-five videos of chimpanzees using tools on ant nests; they conclude that the particular chimp groups they observed have "developed a specialized method for preying on army ants, which involves the use of an additional tool for opening nests" (1).

Dipping for ants is a well-established chimpanzee tool-use behavior (Shumaker, Walkup, and Beck 2011). Chimps take some sort of probe (typically a stick) and poke it into a large army ant nest. Ants climb up the stick, and the chimp enjoys an antsicle. Dipping probes vary in length based on the ant or termite species being hunted and on the materials available to the hunting chimpanzees.[1] In the Goualougo Triangle of the Republic of Congo where Sanz and her team perform their research, several groups of chimps preying on several *Dorylus* species of army ants have been found to employ two different types of tools: one is a "woody sapling to perforate the ant nest" and the other is "a herb stem as a dipping tool to harvest ants" (Sanz et al. 2009, 1). This animal use of tool sets is even more rarely witnessed than the use of single tools; with this discovery, Sanz and her team have asserted that these chimps "exhibit more flexibility in order of tool types" than that observed in chimps' honey gathering activities, another documented chimpanzee use of a set of tools (2).

From the artifacts gathered, video clips, and observations, we know that the two tool types in the set are used in concert and are clearly different. The nest-perforating tools have an average length of 92.3 cm and average diameter of 7.3 mm and are typically made from saplings with leafy branches still intact on the far end (4). The ant-dipping tools are composed of a more flexible material and have an average length of 64.4 cm and diameter of 5.4 mm. These ant-dipping tools are made from "herbaceous materials" gathered by uprooting plants; many recovered dipping tools still had roots on the unused ends (4). The two tool types are procured differently and are different in material composition, length, and diameter; they were also observed being used differently. Researchers observed ant nests being perforated with the perforating tools eleven times: the sapling was inserted into the nest, partially withdrawn, and reinserted several times, then withdrawn; the herbaceous dipping tool was then inserted to gather ants onto the tool, making an antsicle (4). Researchers recorded forty-eight bouts with the dipping tool, sometimes being used along with bouncing, twisting and tapping, possibly to "stimulate movement of the ants onto or up the tool" (4).

The nest-perforating tools observed by Sanz and her team are similar to nest-digging tools seen in Guinea; the researchers suggest that "it is possible that this tool set once had a wider distribution or has been independently innovated in different chimpanzee populations across western equatorial Africa" (5). Nest-digging and ant-dipping tools, which have been observed in Guinea on a few occasions, use tools that are different in size and composition. Researchers speculate that different army ant species that live in various areas may react differently to various tools, and that this may contribute to the observed differences in chimpanzee populations' material cultures. In the case of the Goualougo chimps:

> The use of perforating tools instead of hands for opening nests might confer two advantages to the chimpanzees of Goualougo (a) it elicits a less aggressive attack from the ants which may in turn allow overall longer dipping times and higher yields and/or (b) it causes less disturbance and reduces the likelihood of an early migration of the army ant colony so that chimpanzees can continue to exploit the same nest repeatedly over the course of days or weeks. (5–6)

The ability to eat from the same nest over long periods of time may offer the chimpanzees greater nutritional gain. Additionally, the use of nest perforating tools may be better for the structural integrity of army ant nests, "which means that the ants may stay longer in their nest and not quickly migrate to another location in response to the attack" (6). Repeatedly using the same nest both reduces chimpanzees' food search times and "could also be a form of 'sustainably harvesting' this food resource" (6).

These Goualougo chimps also use several other tool sets to "open a substrate and then gather the embedded food resource": puncture and fishing tools, used on subterranean termites and epigaeic termites, and several types of tools used in honey gathering, including thick tools from dead branches used to pound beehives (6). The variety of tool use among these chimps is notable, as is their use of complementary tools in concert with one another.

At another research site in Bossou, Guinea, wild chimpanzees regularly use stones to crack nuts to get to the kernels. This nut-cracking tool use is supposedly the most complicated known tool use by chimps in the wild (though more impressive feats have been caught in laboratory settings). According to Inoue-Nakamura and Matsuzawa (1997),

> It basically consists of the following actions: (a) picking up a nut, (b) putting it on an anvil stone, (c) holding a hammer stone, (d) hitting the nut on the anvil stone with the hammer stone, and (e) picking up the kernel in the cracked hard shell and eating it. (159)

Nut cracking has only been observed in limited areas of West Africa, and not at all in East Africa, although nuts and stones are available in both locations. Inoue-Nakamura and Matsuzawa see nut cracking as "a strong example of the diversity of material culture among chimpanzees" (160). Bossou chimps have tool-use behaviors that include: "nut cracking, ant catching, leaf folding for drinking water, and pestle pounding for sap extraction from oil-palm trees" (160).

Inoue-Nakamura and Matsuzawa performed an experiment about nut cracking with infant chimpanzees to examine how wild chimpanzees develop stone tool use; they discovered that, like human infants younger than two, chimp infants at two and a half years old have basic touch-together actions but cannot combine actions to crack a nut (or perform some human equivalent, such as using a spoon properly). In both these groups, "the lack of adequate composition of each action is a common characteristic" (170). By comparing the development of tool use in young chimpanzees to human infant development, these researchers demonstrate that the learning processes of the two species are not dissimilar. The researchers summarize the learning process of these chimpanzees in the following way:

> True imitation cannot explain the results of the present study. The infants showed various combinations of stones and nuts. They also showed a variety of fundamental actions. Not all of them were adequate actions for nut cracking. They gradually increased the relative frequency of adequate sequence of the basic actions through each stage of development. They did not copy the motor patterns or the way to relate nuts with stones, which were shown in the tool use by mothers and other members of the community. As the present results suggest, they learned the general functional relations of stones and nuts and also learned the goals obtained by the demonstrator. This learning process might be called emulation. (172)

The chimpanzees' learning process does not precisely mirror the human case:

> Active teaching seems to be a clear distinction in the process between humans and chimpanzees. Active teaching is rare in chimpanzees in the wild although it is popular in humans. So far, there have been no instances of active teaching or guidance except two reported episodes in which mothers influenced their infants' attempts to crack nuts. (172)

The actual variety and cultural differences observed in chimpanzees' learning processes in the wild indicate that we greatly underestimate the sorts of work being done outside the human domain.

Kanzi

Like Koko the gorilla, who successfully learned sign language, Kanzi the bonobo chimpanzee is a famous subject of research into the abilities and capacities of great apes. Susana Carvalho, Tetsuro Matsuzawa, and William McGrew (2011) summarize Kanzi's impact as follows:

> The Kanzi study showed that the bonobo, a non-technical species in the wild, had the necessary cognitive flexibility and motor coordination to solve a problem that requires the making and using of a stone tool. This research also suggests that during learning, innovation and idiosyncrasy may have played crucial roles in the emergence and variability of technology. (227)

What I describe here comes from the research of Nicholas Toth and his colleagues. Though less famous than Koko, Kanzi's learned tool use and innovative techniques point to apes' having a greater capacity for and flexibility in tool use than previously assumed. Kanzi, a bonobo chimp used mostly in language studies, also took part in an important study on tool use.[2] Researchers set out to see if bonobos could learn to make stone tools, like those made and used by humans' hominid ancestors (Schick and Toth 1993, 135). Based on the work of Richard Wright (an archeologist who demonstrated in the 1970s that an orangutan could be taught to make stone flakes to use as a tool to cut a string on a box containing food), and using Oldowan stone tools as a guide, researchers first showed Kanzi how useful stone tools could be by cutting cords to open boxes with Kanzi's favorite treats in them. On the first day, Kanzi used the stone flakes made by researchers to cut into boxes. On the second day, Kanzi could judge the best tool for the job, choosing the sharpest knife most of the time; Kanzi was also hitting stones together to make his own tools (Carvalho, Matsuzawa, and McGrew 2011, 136). Within a month of the start of the study, Kanzi was striking rocks to make flakes with hard-hammer percussion. He was given some demonstrations by researchers of how modern humans make tools (136); within several months, Kanzi came up with his own technique—throwing a stone on his hard floor to splinter them into edges for cutting. The researchers note parenthetically, "Curiously, he had never been very interested in throwing until this technological advance of his, although some of his siblings have shown a penchant for throwing things. But now he seemed to find a purpose to it" (138). The researchers refer to this new process of Kanzi's (repeated floor throwing) as an innovation: Kanzi "seemed to have made his own connections between the force of the throw and successful flaking" (138).[3]

With this new association, Kanzi's hard-hammer percussion work "became much more forceful," and Kanzi started producing better flakes with this method, "producing objects that we can begin to compare with those in the early archaeological record" (139).

Researchers note that after nine months, Kanzi's work was still not of comparable quality to the found Oldowan tools that the researchers used for comparison, but the researchers were impressed by his innovative technique and the new understandings he displayed as he worked with the tools (139). Though this study was conducted in captivity, it demonstrates bonobos' capacity for tool use, speaking against some philosophers of technology and anthropologists who insist that humans are the only species capable of innovating in their technologies and techniques. Kanzi demonstrated innovation in tool use when his new throwing technique led to better hard-hammer percussion of stones and produced better stone tools.[4]

In studies that followed, capuchin monkeys in captivity produced stone flakes to use as cutting tools and exhibited the patterns of right-handedness that some associate with tool-making hominids. Hominids produced the first stone tools 2.5 million years ago in what is called the Oldowan technological stage (Schick and Toth 1993, Westergaard 1995); anthropologist Louis Leakey first discovered these tools, which consisted of flakes and battered stones, in an area of Tanzania during the 1930s. The external objects employed in the documented tool use of wild chimps comes largely from vegetation—bent twigs, leaves, sticks—although some chimps use stones to crack open nuts (Westergaard 1995). Capuchin monkeys also produce tools in a series of contexts: clubbing snakes with sticks, using shells to crack open oysters, etc. Capuchins who are, like Kanzi, taught to flake stones in captivity demonstrate "impressive artifact production skills" (Westergaard 1995, 3). Capuchins also demonstrate right-handedness at a population level, something that some researchers associate with the complex object manipulation that can be related to the evolution of language (Westergaard 1995).

Great apes (humans, chimps, orangutans, and gorillas) are more closely akin than other primates. The category of *ape,* in its current understanding, includes humans and the non-human species of chimpanzees (bonobos and common), orangutans, and gorillas, as well as the "lesser apes," a category which includes the sixteen known types of gibbons. Apes are a subset of primates, where the category primates includes New World monkeys and Old World monkeys, as well as lemurs and several other species. Not a type of ape but still in the primate family, capuchin monkeys are a type of New World monkey. The capacity for the use and production of tools certainly extends beyond the closest human relations, as demonstrated by the Capuchin monkey studies in tool use.

Santino and Chimp Planning

My favorite chimp case comes neither from a laboratory nor the field, but from a zoo. In the popular press, it was reported that a chimpanzee at Furuvik Zoo in Sweden stockpiled rocks to pelt at zoo-goers. This chimp, Santino, furtively stashed away stones, demonstrating "the first unambiguous evidence of spontaneous forward planning in a non-human animal" (Henderson 2009). According to Mathias Osvath, who published the study of Santino,

> Planning for a future, rather than a current, mental state is a cognitive process generally viewed as uniquely human. . . . [Santino's] planning actions, which took place in a calm state, included stone caching and the manufacture of discs from concrete, objects later used as missiles against zoo visitors during agitated chimpanzee dominance displays. Such planning implies advanced consciousness and cognition traditionally not associated with nonhuman animals. (Osvath 2009, R190)

Santino's stone-caching behavior was not a one-time occurrence. In June of 1997, Santino threw stones several times in his aggressive displays; zookeepers did a sweep of his enclosure, finding five caches with three to eight stones each, as well as individual stones between caches, all along the shore facing the public (ibid., R190). On five consecutive mornings, a hidden zookeeper watched the chimp gather stones from the water and cache them. A year later, in June of 1998, Santino added concrete pieces to his caches, breaking off chunks from the large concrete rocks in the center of the island enclosure (ibid., R191). Santino even had a technique for making these concrete rocks to hurl at visitors: he "was observed to gently knock on the concrete rocks from time to time, delivering harder blows to break off the detached surface section in discoidal pieces, and sometimes breaking these into further smaller fragments" (R191). Since 1997, zookeepers have removed hundreds of Santino's caches, though Santino never found, stored, or made these missiles while zookeepers or visitors were obviously present. In the off-season and off-hours, Santino does not throw stones, and his stone caches are located only along the shoreline that faces the visitors' area, making the purpose of his stone caching pretty apparent (R191). Santino was certainly never taught how to use or make the concrete discs he throws at visitors; it seems he came up with their production and use on his own. (No zoo-goers have been hurt, as Santino only throws underhanded; Henderson 2009).

Social Transmission and Cultural Evolution

Recent studies bolster the idea that, among great apes, tool behaviors are culturally transmitted; chimps are especially capable of learning and of inventing new techniques. At Kyoto University's Primate Research

Institute, researchers Shinya Yamamoto, Tatyana Humle, and Masayuk Tanaka gathered evidence in experimental settings about social transmission of a tool use techniques. They write:

> Many researchers . . . consider that only humans are cognitively capable of cumulative cultural evolution. However, experimental conditions or motivational factors may undermine the chimpanzees' abilities and performance. Here we present the first experimental evidence, to our knowledge, for chimpanzees' social learning of a more efficient tool-use technique in an intuitive tool-use situation, suggesting that the limitation for chimpanzees' cumulative cultural evolution might be due to ecological, social, and motivational factors rather than cognitive inabilities per se. (Yamamoto et al. 2013)

While studies of chimpanzees and bonobos in the wild have pointed to some of these findings, the Yamamoto team tries to show how chimpanzees in captivity use social learning, which could suggest cumulative cultural evolution. Researchers provided a task in which their nine chimpanzee participants were given a straw to obtain juice through a small hole. Four of the chimps sucked from the straw ("the straw-sucking technique") and five of the chimps used a less efficient "dipping technique." Straw-suckers were able to get more juice more quickly. The researchers gave the participants this test separately over five days, and those using the dipping technique failed to innovate in this time, continuing to use their dipping technique.

Next the Yamamoto team paired chimpanzees who used the dipping technique with chimpanzees who using the straw-sucking technique and gave them the task. They describe the results:

> When we paired each of the five "dipping" participants with a "straw-sucking" conspecific non-kin demonstrator in the same booth, four of the five participants subsequently adopted the "straw-sucking" technique. Those chimpanzees who most closely and attentively observed the demonstrator perform the alternate "straw-sucking" technique switched more rapidly to using the novel technique . . . Mari, who never closely attended conspecific demonstrations of the "straw-sucking" technique and consequently never learned this alternate technique, only finally switched her technique after watching consecutive demonstrations performed by a familiar human. Once the chimpanzees switched to using the "straw-sucking" technique, they never again reverted to using the less efficient "dipping" technique. (Yamamoto et al. 2013)

The researchers observe that necessity and opportunity play key roles in determining technique change. This study and others suggest that chimpanzees who do not switch their techniques are not less cognitively able than their technique-switching counterparts; instead, chimpanzees choose to switch techniques only when they are not sufficiently pleased with their current techniques (Yamamoto et al. 2013). Testing the chimpanzees as the Yamamoto group did—with a wild-like tool use situation that did not require varied experimental set-ups—preserved the chimpanzees' own original techniques, and documented how other chimps responded to observing a better technique; this helped to isolate the variables.

Isolating factors to gather clear data about social learning can be very difficult. One team of researchers used network-based diffusion analysis, a form of social network analysis, to examine patterns in the spread of a technique in an effort to discover whether technological behaviors could be explained by social factors (Hobaiter, Poisot, Zuberbuhler, Hoppitt, and Gruber 2014). These researchers analyzed observations from the Sonso chimpanzee community of Budongo Forest, Uganda, where they examined the use and spread of two tool-use variants. Researchers concluded:

> We find strong evidence for social transmission of "moss-sponging" (the production of a sponge consisting of moss) along the innovators' social network, demonstrating that wild chimpanzees learn novel tool-use behaviors from each other and supporting the more general claim that some of the observed behavioral diversity in wild chimpanzees should be interpreted as "cultural." (Hobater et al. 2014)

The network analysis technique provides more evidence for chimpanzees' social learning (and for cultural transmission). Taken alongside the other evidence provided about chimpanzee cognition and tool-making, this study bolsters the claim that many of the factors we consider definitional in the study of human tool use are also present in other great ape species.

COMPARATIVE STUDIES

While the last section looked at a few notable chimp cases, comparative work about tool use that indicates differences between species and cultural differences also helps make a case that an understanding of tool use is part of the lives of great apes.

Chimps, Bonobos, Orangutans, and Capuchins

In 1995, Elisabetta Visalberghi, Dorothy M. Fragaszy, and Sue Savage-Rumbaugh (a team member in the Kanzi studies and an important figure at the Georgia State University Language Research Center) published a comparative study on tool use in bonobos, common chimps, orangutans, and capuchin monkeys. They noted differences between the great apes in the study (bonobos, chimps, orangutans) and the monkeys (monkeys are not apes). The study included six capuchin monkeys, five common chimps, four bonobo chimps, and one orangutan (and a partridge in a pear tree), all of which were born in captivity except one of the bonobos (Visalberghi et al. 1995, 53). None of the primate participants had encountered this particular experimental set-up, though a number had participated in other tool-related experiments.[5] The participants ranged in age from 2.5 to 20 years old (54). The great apes in the study—the chimps and orangutan—participated from the Language Research Center, while the monkeys encountered the similar experimental conditions at Washington State University (53–54). The relevant experimental apparatus consisted of a clear tube with a treat inside and different tools for treat extraction. In the first trials of the experiment, the tool available to the test subject was a wooden dowel with which the participant to could push the treat out of the tube, gaining access to the "highly preferred food" (54). After mastery of this setup was achieved (for the monkeys, this meant ten consecutive successful trials; for the apes, this meant successful performance for three consecutive trials), the participants were given more complex conditions: two tools, a bundle of dowels, and an H-shaped stick constructed from dowels, placed much further away from the tube (54). For the bundle of dowels, the participants had to take apart the bundle to use one dowel in the tube for successful completion of the task. For the H-shaped sticks, the H had to be deconstructed so that one line of the H could be used to extract the treat from the tube.

Four of six capuchins solved the simple setup on the first trial, and the other two succeeded by the fourth time (55). Eight of ten apes solved the simple setup on the first trial; the two youngest participants did not work as quickly, with one solving on the seventh trial while the other was removed from testing after thirteen trials (this ape was tested again two years later).[6] All but one of the monkeys made errors in the complex bundle trials, attempting to insert the bundle before taking the bundle apart, although they did eventually succeed. In contrast, the apes solved the complex bundle problem easily, with none of them trying to insert the whole bundle. All of the monkeys made errors on the H-shaped complex trial (55): they inserted the unbroken H and inserted the short cross-dowel that held together the two long dowels (the long dowels were the ones that should be used to get the treats). The apes made three types of errors on the H-shaped complex trials, but performed much better, with researchers concluding that "[o]verall it appears that apes have or easily acquire an appreciation of the length of the stick needed for solution" (56). Capuchin monkeys were found to succeed just as often as apes, but with more inefficiency in subsequent attempts and new situations; the authors explain that these trials demonstrate that tool use is not simply "a yes-or-no capacity," and that equal success at a task may not "imply equivalent understanding" (56):

> The persistence of errors across blocks of trials suggest that the subjects did not achieve full comprehension of the requirements of the task within the testing period . . . their behavior toward the task was not random, and it improved (although not significantly) with experience. (58)

All three ape species tested performed similarly well; apes did much better on subsequent trials than the capuchins, with none of the apes trying to insert the bundle of dowels. While capuchins continued to make mistakes, the ape participants in this study improved with practice. The researchers conclude that:

> These findings suggest, at the least, that apes can acquire more readily those associations relevant to solution of the task. An equally plausible interpretation is that apes acquire a fuller comprehension of the task than capuchins do. . . . [A]pes and capuchins can achieve success in a tool-using tasks and still have an apparently limited understanding of the causal relations involved . . . It appears that the experience of using tools does not, by itself, lead to or require the emergence of additional conceptual complexity. Nevertheless, apes appear to make greater progress to this end than do capuchins. (58–59)

The researchers tested understanding of the causal relation between a tool-using act and its outcome, and found differences in the methodologies and success rates of monkeys and apes. Of course, further research on gorillas (a great ape that was not included in the study), gibbons (lesser apes), and other types of primates are needed before making definitive category distinctions between primates based on tool use.

Interestingly, this study observed a noticeable difference in ape performance by age, but no difference in capuchin performance by age. The youngest capuchin, who was 2.5 years old, performed as well as the other monkeys; the two youngest apes were the least successful at the outset (58). This finding suggests that apes and monkeys have different learning and development trajectories, though more research would be needed to draw any conclusions.

Conceptual Categories

Comparative studies of how different groups of primates use tools may tell us something about these groups' differing capacities or environments, but intra-species studies yield intriguing results about cultural differences among or in animal groups. There is evidence of material cultures among several species of wild animals, including wild chimpanzees.

A foundational study that compared tool use among three wild chimp populations demonstrates differences between these populations' tool making, tool use, and tool diversity. Boesch and Boesch define tools as objects that "must be held in the hand, foot or mouth and used as to enable the operator to attain an immediate goal" (1990). Using data from three long-term studies of wild chimp populations in East and West Africa, the researchers discuss the impact of environment on tool-related behaviors. There were three populations examined: one from a rainforest environment on the Ivory Coast, West Africa (Tai National Park), and two from parks in a savannah/woodland environment in Tanzania, East Africa (Mahale Mountains National Park and Gombe Stream National Park).

During a nine-year period, researchers in the Tai National Park observed the local chimpanzees using tools for five different activities (insertion, probing, cleaning, displaying, and pounding) for eighteen aims (including ant-dipping, honey fishing, boring bee nests, sponging off, throwing, dragging, etc.). In addition to ant-dipping (the most common chimpanzee tool-use type), for example, Tai chimps are known to use tools for hammering open nuts and eating bone marrow (and other parts) of colobus monkeys (Boesch and Boesch 1990, 89–91). Chimps in the Mahale Mountains and Gombe Parks seem to have only four different tool-use activities, and their tool-related aims differ widely, with Gombe and Mahale chimps doing more tool-aided cleaning and using insertion and probe tools in fewer contexts.

Tasks performed by both rainforest and savannah chimps (such as ant dipping) are carried out using different strategies in the different environments. Researchers observed Tai chimps using their hands to take eggs, larvae, and pupae out of an ant nest, and to excavate the nest, inserting their arms to the shoulder; they also use short sticks, averaging 23.9 cm in length, for ant-dipping. They let ants crawl 10 cm up the antsicle before eating and dipping again (89), and they dip around 12 times per minute, with an average of 15 ants per dip (89). Gombe chimps use ant-dipping sticks of around 66 cm and dip 2.6 times per minute, with 292 ants per dip (91); they are less likely to directly remove grubs from nests with their hands and more likely to rely exclusively on tools in ant-hunting (91). Mahale chimps sometimes fish for tree ants with tools, but have not been seen to ant-dip like Gombe and Tai chimps, although the Mahale chimps are geographically close to the Gombe and share an environment with them (92).

The behaviors of these groups of chimps do differ by location, of course. The savannah chimps, the Gombe and Mahale groups, "seem rather fastidious" and researchers have observed that "males regularly wipe semen from their penis after copulation, a behavior never observed at Tai" (93). And their tool-use patterns also betray geographical differences: the rainforest chimps at Tai have been observed performing 19 types of tool use, while the savannah chimps, at Gombe and at Mahale, have been observed performing 16 and 12 types, respectively (93). And only the Tai chimps have been observed "to pound objects with tools and to combine different tool uses to get access to a single food item" (93).[7] The norms for tool use and for hygiene differ between these populations of the same species in different locations.

There are also striking differences in the tool-making procedures of these chimp groups. The very methodology of tool-making differs:

> Qualitative descriptions from Mahale indicate that, when making a tool, the chimpanzees tend to modify it progressively, i.e. testing the tool after each modification until it becomes adequate. Thus, the standardization of the tools in Mahale chimpanzees seems to be only the result of the successive improvements made on the tool during use. In contrast, . . . Tai chimpanzees proceed to all modifications before using the tool. Hence, tool making in Tai chimpanzees seems to require a precise idea of the form of an object must have to be considered a tool, as well as all of the technical steps necessary to perform on it to conform to this predefined idea. (95)

This difference, not only in use and aim of a tool but in the process of making it, points to a sharp contrast in these chimpanzee communities' material cultures and practices.

Research indicates that early learning affects what tool behaviors a group of chimps pass on to others in their group. For example, the Tai chimps are the only chimps observed to use hammers to crack nuts, although there are hard-shelled fruits and hard objects in Gombe that would invite this behavior. However, "such demanding tool use can develop only in a situation of rich nutritional rewards," and young chimps learn what to eat based on "food sharing in the family" (6). Researchers have observed "[t]he more prominent role of food sharing during meat-eating episodes" in Tai chimps, which encourages this group's acquisition of nut-cracking skills. Environmental factors, like materials available, also influence the sort of tool use of a group:

> Tai chimps have apparently developed high faculties in representation of space for finding the rare stones at previous nut-cracking sites rather than developing sophisticated techniques to make tools of hard material: raw material being rare, it is more economic to search for tools rather than to search for raw material. . . . Tool making in chimpanzees seems to be inversely proportional to the availability of material. (97)

Selectivity, the making of tools, and the availability of raw materials are thus related; according to Boesch and Boesch, "an increase in the sophistication of tool making may permit less selectivity for the raw material and individuals become less dependent on the environment" (97).

There have been similar studies on orangutans' tool-use behaviors indicating that cultural variation in tool use among dispersed orangutan species cannot be simply explained by phenotypic variation; it also depends upon environmental and developmental differences (Krutzen, Willems, van Schaik 2011).

CONCLUSIONS

The tool use of chimps, orangutans, and monkeys covers a range of technological behaviors, though most tools are used in service of procuring food. The available data on the tool use of great apes indicates several facts:

- At least some ape groups have material cultures, which differ by population, place, prey, availability of raw materials, learned traditions of tool use, etc.
- Chimps, if not other types of apes, are capable of standardization in the making of tools.
- Chimps use a variety of tools, sometimes in concert, and these tools vary by population.
- Chimps, at least in captivity (although there are strong examples from the wild that I have not covered here), have the *capacity* for innovation. Kanzi's new idea and the understanding shown by apes in the comparative study point to this conclusion.
- The young life of an ape is extremely important to their future tool use and comprehension. This is further confirmed by the language studies on Kanzi, Koko, and other primates. Though I have not described the Koko studies here, they are readily found via an internet search.
- Chimps (and perhaps other primates) can unambiguously plan for the future. The example of Santino and his rocks at the zoo should be the beginning of serious research into chimps' planning abilities.[8]

Especially in more complex environments, wild chimps do plan in advance. It seems that we need more research to be able to discuss the lifeworlds of chimpanzees in a richer way. As Osvath explains, "They most probably have an 'inner world' like we have when reviewing past episodes of our lives or thinking of days to come" (2009).

The tool use of chimpanzees and great apes (and possibly of other primates) is not separate from humans' uses of tools and technology. In fact, many of these studies compare these animals' tool use to that of human toddlers, who have a similar capacity to successfully complete technological tasks (Visalberghi et al. 1995; Inoue-Nakamura and Matsuzawa 1997). As anyone who has ever watched a toddler struggling to insert a block into a shape sorter might attest, the problem displayed in the comparative learning of apes and capuchins is not necessarily a problem of understanding; even when understanding is present, the physical capacity to perform some tool-related tasks must be learned.[9]

These studies can show us how to make useful distinctions between species that might appear on our emerging tool-use map. While capuchins' tool-related processes and understanding differ from those of the great apes and from those of human beings, these should not be ranked hierarchically—recall that the capuchins succeeded as often as the apes, and though their problem-solving approach took longer, their methodology may be the more successful one in the capuchins' home environments.

My goal in this book, however, is not simply to change the human clause to "the human and chimpanzee clause." My project is much more ambitious, and chimps are just the first stop. I hope not just to expand the human clause but to blow it up. It is easy to accept primate understanding and tool use, as they are our "younger brothers in knowledge"[10] and members of our wider family in popular cultural narratives. We feel understanding for these animals more easily than for other species—we understand their facial expressions and body movements better than we do those of more distantly related species. Chapter 5 is on dolphin tool use, and chapter 6 is about birds; accepting these animals' tool use does much more to speak against the human clause, for it sets up discussions of construction and technical skill in a much wider context.

NOTES

1. This cultural difference in ant-dipping will be described in more detail below in the section on Cultural Differences.

2. Kanzi is best known for his participation in language studies with the Language Research Center at Georgia State University. Raised in captivity and exposed to language experiments on his adoptive mother, his ability to use and manipulate language has demonstrated the ability of bonobos to pick up language through social interaction. The best reading on this topic may be Savage-Rumbaugh, S., and R. Lewin. 1994. *Kanzi: The Ape at the Brink of the Human Mind.* Wiley.

3. Describing chimpanzee behavior as innovative is not new. H. Kummer and Jane Goodall have described the "Conditions of innovative behavior in primates" in *the Philosophical Transactions of the Royal Society of London B* in 1985 (volume 308, pages 203–214). To Kummer and Goodall, an innovation may be "a solution to a novel problem, or a novel solution to an old one; a communication signal not observed in other individuals in the group (at least at that time) or an existing signal used for a new purpose; a new ecological discovery such as a food item not previously part of the diet group" (page 205).

4. Incremental improvements like this may even point to the sort of feedback loops that Joseph C. Pitt promotes as an important component of technology (2000), even if his definition of "humanity at work" for technology leaves little room for our animal cases.

5. In fact, Kanzi is one of the participants in this study, as well as his adoptive mother Matata.

6. This youngest ape, Tamuli, age 3, would be subjected to this experiment again 2 years later and perform successfully. The youngest monkey in the study was 2.5.

7. Although we also know now that tool sets are used by Crickett Sanz's Congo chimps.

8. While we may not be able to generalize about what wild chimps *do* from the study of a single captive animal, we can definitively say something about the *capacities* of the species for certain tasks—even mental tasks, like planning for the future. Santi-

no could very well be Chimp Einstein and not indicate much about chimps in general, but his behaviors do demonstrate that such planning is at least possible for chimps.

9. There has been work done on the "specialness" of the human hand in allowing for the ability to make and manipulate tools—it might be that humans are better suited to a wider variety of environments and tasks because of our hands, and this might also explain the extreme variety and endemic nature of human tool making and use.

10. This phrase is from John Sargeant in 1697 in talking about "brutes," meaning animals more generally, but I think this phrase illustrates feelings people have about primates rather well.

FIVE

Cetaceans

WHY CETACEANS?

There are several characteristics that we see as uniquely human and as leading to the rise of technologies: planning for the future, a material culture, social transmission of knowledge, and ability to innovate. As we have seen, great apes clearly display these behaviors. But while the great ape cases of tool use are startling, they are not very controversial—we humans already think of primates as being somewhat like us, with their opposable thumbs and dexterity (both of which are seen as important in tool work). Simply replacing the human clause with a primate clause does not go far enough, as the non-primate animal cases examined in this chapter and those that follow indicate. In this chapter, I propose that cetaceans also be included on our developing map of tool use and technological behaviors, both because of their demonstrated cognitive and linguistic abilities of manipulation and because a few cases of cetacean tool use have been witnessed in observational studies. Cetaceans certainly have techniques, and a few tools, though we may question whether they have the ability to manipulate and form material objects like apes do (and, as we'll see in the next chapter, like crows do).

The category cetaceans includes whales, porpoises, and dolphins. In this chapter, I focus first on the dolphin literature, then offer findings from whale research. (There is little useful literature on porpoises and tool use.) The literature on dolphin tool use is much less extensive than that on chimpanzees, in part because there are increased barriers to observation and collection for ocean mammals, and in part because dolphins have been the object of this type of study for only a few years; although observation of chimp tool use goes back to 1960, dolphin tool-use research lags far behind—the first case of dolphin tool use in the wild was reported in 2005. Therefore, the literature on cetacean tool use reported in this section will necessarily be somewhat less comprehensive. I do, however, offer as a supplement research into dolphins' and whales' cognition and linguistic abilities. Language is also a tool.[1] Linguistic abilities can be considered to be either parallel to technological capability or groundwork for technological capability, and I treat bottlenose dolphins' ability to manipulate and use sounds both as language and as a technological endeavor.[2] Similarly, particular cognitive abilities subtend the defining characteristics of tool-using animals (planning for the future, the social behaviors that create shared material cultures and allow for social transmission of knowledge, and ability to innovate), and I therefore examine studies into cetacean cognition here. After all, aren't these cognitive and linguistic capabilities what the human clause is *really* trying to get at? Although the human clause tells us that humans are special because of our use of technology, technology is, in this definition, conflated with and related to intelligence and linguistic ability—our ability to transfer things *and* ideas to one another.

SPONGING AND SOCIAL TRANSMISSION

If dolphins' linguistic feats are set aside, examples of dolphin tool use are few, but this is, in part, because researchers did not begin to look for such examples until recently. Why would they look for something thought impossible? It is difficult to imagine that dolphins and other marine animals could fashion tools; so many accounts of tool use describe the importance of hands to manipulating and wielding. However, in Western Australia's Shark Bay in 2005, a population of wild bottlenose dolphins that had been under observation since 1984 displayed eleven foraging tactics, "exhibiting a diversity comparable with that of chimpanzees and orangutans" (Krützen et al. 2005, 8939). Only one of these tactics involves the use of a tool, and this tool use has been witnessed in only fifteen of 141 known dolphin mothers and in seven of their young. This dolphin tool-use behavior, which has been termed "sponging," consists of breaking off a piece of marine sponge from the seafloor, placing it over one's nose (rostrum), and using it to probe the seafloor for hidden fish. Sponging is almost exclusively used by females as a foraging technique; all but one of the observed spongers were female.

Sponging fits classic definitions of tool use, such as Benjamin Beck's: "the external employment of an unattached environmental object to alter more efficiently the form, position, or condition of another object, another organism, or the user itself when the user holds or carries the tool during or just prior to use and is responsible for the proper and effective orientation of the tool" (Beck 1980, 10). In this instance, the observed dolphins detached an environmental object (the sponge) to position it in such a way to carry it and orient it to effective action in scrounging up fish.

Researchers rule out ecological factors as the sole motivation for this sponging technique, because other female and male dolphins in the same areas were seen to forage without using the technique (Krützen et al. 2005, 8939). The researchers think it seems "highly likely that sponging is culturally transmitted mainly within a matriline, i.e., daughters learn this behavior from their mothers" (8939). To rule out genetic influence on the transmission of this technique, and to examine their theory that the trait is culturally transmitted, researchers did genetic and statistical analysis on spongers and nonspongers. According to the researchers, "previous genetic analyses showed that random mating can be assumed for the Shark Bay population. Hence, if the relatedness levels among all spongers were significantly above the population average, then it would be likely that sponging was a fairly recent invention" (8940).

The genetic study concluded that the spongers were indeed "closely related to each other and share a recent common ancestor" (8942), with relatedness levels that were indeed significantly above the population average; this implied both that cultural transmission was likely (8943) and that sponging was indeed a recent invention (8940). This study drew interesting headlines in the popular press about dolphin mums teaching their daughters a new technique.

Since the discovery of sponging, researchers have observed more than fifty dolphins (about 5% of the dolphin population in Shark Bay) using the sponging technique (Mann and Patterson 2013). Shark Bay dolphins have also been found to use other objects in ways that might, after further study, be classified as tool use; these informally observed behaviors include juveniles playing with sea grass and surfacing with large gastropod shells (Mann and Patterson 2013). Dolphin foraging techniques, like chimpanzee tool use and making, have been documented to vary by population, even in similar environments (Sargeant et al. 2005). When genetic research rules out shared biological or instinctive bases for these techniques, the behaviors that remain almost certainly indicate shared cultural or social practices and techniques. Citing a variety of studies on bottlenose dolphins, B. L. Sargeant et al. report that bottlenose dolphins

> forage both in groups and individually and have also adapted to human activity by following fishing boats to obtain discarded fish, visiting provisioning locations, and catching fish cooperatively with net fishers. Additional tactics include using their rostra [noses] to dig into the substrate, smacking their tails on the water surface over shallow seagrass beds to disturb prey, whacking fish with their tails, foraging with the aid of marine sponges worn over their rostra, and stirring up sediment to trap fish, among other behaviors. (1401, in-text citations removed)

In Shark Bay, the same research site where the sponging technique was observed, researchers conducted a longitudinal study of mothers and calves that documented thirteen foraging tactics. Different dolphins specialize in different tactics; researchers observed one particular tactic, beach hunting, used by only "a handful of known individuals," and performed a case study of the technique's development and how it was taught and/or learned. In beach hunting, one dolphin "surg[es] partially or fully out of the water and onto the beach to catch a single fish" (1401). This tactic, which is easily observed, has similarities to other cetacean hunting tactics: killer whales beaching themselves to catch pinnipeds in Argentina; humpback dolphins at low tide pushing fish up onto exposed sand near the Bazaruto Archipelago; bottlenose dolphins beaching fish on mudflats off the southeastern coast of the United States, etc. (1401). With these types of beaching tactics, however, there is always a risk that the hunter may become stranded on the beach. In this case study of beach hunting, researchers observing East Peron Beach (often used by beach hunters and easy to observe) identified individual dolphins by fin shape and markings. Observations were taken from the water (in a boat) and from the shore. Over 96 hours of observing already identified dolphins, 98 beach-hunting bouts by four beach hunters and their juveniles and calves were observed; 28 of these involved a full beaching onto shore (1404). The identified prey included mullets and longtoms, and, in every case observed, the dolphin went after only one fish at a time. Calves and juveniles were seen engaging in the method, but none of them used full beaching as a technique. There was no coordinated or cooperative foraging by the beach hunting method, though some dolphins pursued different individual fish within meters of each other (1404).

The researchers in Cape Peron wanted to figure out whether teaching—active instruction, not simply a chance for observational learning—played a role in young dolphins' acquisition of this technique. Younger calves were found to wait apart from their mothers, observing beach hunting from further away than older calves and juveniles (Sargeant et al. 2005, 1405). However, "mothers were never observed attending to their calves or altering their foraging behavior in response to calf presence during beach hunting, presenting no evidence for teaching" (1405). With the evidence on hand, it seems that dolphin calves learn much like young chimps: techniques appear to be learned by observation of their own mothers' behavior. The beach hunting technique does seem to be passed mother to daughter, however. Although the two most frequent beach hunters associated with at least 25 other individual dolphins (not including their own young and other known beach hunters), and spent frequent, regular time with two other particular dolphins over the past 10 years, none of these other dolphins, including the two frequent associates of the beach hunters, have ever been observed to use the beach hunting technique. As the researchers put it, "despite consistent association with many other dolphins, regular use of beach hunting in the Cape Peron area currently appears restricted to just four adult females and their offspring" (1405). Researchers also note that the mitochondrial haplotypes among the beach hunting dolphins differ; three of the beach hunters shared one haplotype (the most common among this particular group of dolphins), and the most frequent beach hunter had a different haplotype (1407).

Using the small sample group of beach hunters, the researchers quantified the amount of time spent by individual dolphins on different foraging techniques, and found that dolphins specialize in foraging techniques, even in large populations that use a wide array of techniques. This means that there are many techniques a dolphin could engage in, but they tend to choose one and stick with it. They also found evidence that "foraging complexity and learning correspond to longer periods of dependency" of young on their mothers; juveniles under the age of five were not observed to try full beaching at all (1407–1408). Although there are examples of possible teaching in chimpanzees and killer whales, researchers found that bottlenose dolphin mothers did not change their techniques in the presence of their calves at all—an action that would have demonstrated that the mothers were teaching their young. In these dolphins, at least, observational, rather than instructional, social learning in the matriline plays a significant role in the acquisition of the beach hunting technique.

SPECIALIZATION AND VOCALIZATION IN BOTTLENOSE DOLPHINS

Sargeant et al. showed that individual dolphins' specialization in foraging techniques is quite strong (2005). However, cetaceans do not always hunt individually; many participate in cooperative or group hunting, including killer whales and humpback whales (Gazda et al. 2005). An observational study was conducted in Cedar Key, Florida, among two groups of bottlenose dolphins, one at Seahorse Key and the other at Corrigan Reef; researchers observed a group hunting method that relied on division of labor, with individual dolphins taking on specialized roles. Group-hunting dolphins in Cedar Key use two behaviors to herd fish for prey:

> One individual in a group of three to six dolphins, the "driver," herds the fishes in circles, as well as towards the tightly grouped "barrier," or "non-driving" dolphins that are less than one body-length apart and often touching. The driver may perform fluke slaps (when a dolphin lifts its fluke, or tail, out of the water and slaps it against the water surface forcefully) during the drive. Fishes being herded in this fashion

leap into the air, where some are captured by driver and barrier dolphins. (135)

In the two groups observed, the herding behavior described above was witnessed and analyzed in 126 cases: sixty from the Seahorse Key group and sixty-six from the Corrigan Reef group (no group members overlapped) (137). In all of the Seahorse Key group hunting bouts, the same dolphin served as the driver; the Corrigan Reef group also had a driver dolphin that performed this job each time (135). There was also one dolphin in the Corrigan Reef group who seemed to practice driving when other group members were not present; this dolphin was a non-driving member in the group hunt (135).

Though cooperative fishing has been witnessed elsewhere, the level of specialization these dolphin groups display in this activity was surprising: "the consistent role-playing in cooperative herding that was seen in Cedar Key" had not been found among other cooperative groups at that point (138). The researchers argued against an interpretation of the results whereby the non-drivers were actually scroungers benefiting from the driving dolphin's work—a "noncooperative" explanation. Average driver success (in terms of fish eaten) did not differ from that of non-driver success: everyone in the group that used this technique was, on average, equally successful (138). They argue instead that the barrier dolphins also play a role in this technique (trapping the fish), and that the driver seems to intentionally move toward the barrier dolphins—an unlikely action if the barrier dolphins were simply scrounging (138). Evidence of this role specialization in group-hunting points toward a sophisticated group dynamic, and these groups are indeed cohesive: they were observed to move "slowly in a cohesive manner along the channels during the search for fish schools," and group members were repeatedly seen together four to five years before this particular study was begun (138).

These dolphin groups' cohesion and complex group dynamics are further evidenced by their vocal learning. Bottlenose dolphins are very social, with pairs of dolphin males and mother-young pairs being observed together "continuously for 5 to 10 years or more" (Tyack 2000). Studies have shown that wild bottlenose dolphins can learn and imitate the signature vocal whistles of their group members (Tyack 2000, Janik 2000, Janik et al. 2006). To study these whistle-matching interactions—interactions in which one dolphin copies the vocal whistle of another, perhaps indicating some form of address—researchers recorded whistle-matching off the shore of Kessock Channel in Moray Firth, Scotland (Janik 2000). In this study, two whistles were said to be matched when they came within 3 seconds of each other from locations more than 26 meters apart and were judged to be matched by "five naïve human observers" who rated the similarity of the whistles (Janik 2000, 1355). Recording equipment was set up to passively record; there were no humans or boats present in the channel during the study (1355). Because individual dolphins' whistles could not be identified at distances closer than 26 meters, researchers note that it is possible that more matching whistle interactions occurred than were reported in this study. Matching whistles were found on all seven days of observation, and typically took place between two dolphins; twice, whistle-matching occurred between three dolphins, and three times, one dolphin gave a whistle that was followed by a matching whistle from another dolphin and then repeated back by the first dolphin (1356). The researchers concluded that "bottlenose dolphins use their learned whistles in matching interactions, most likely to address each other," and that "although vocal matching is common in birds, bottlenose dolphins are the only nonhuman mammals in which matching interactions with learned signal types have been found" (1357).[3] Dolphins' whistle-matching is "thought to function in group cohesion and individual recognition"; indeed, captive dolphins are known to imitate human sounds, something not even apes can do (Janik 2000, 1355; Tyack 2000).

In a follow-up study, conducted in Sarasota Bay, Florida, dolphins were tested to see whether they could identify an individual dolphin without hearing that dolphin's whistle; if so, this would indicate that signature whistles are "independent of voice features, as it is in human naming" (Janik et al. 2006, 8293). The researchers produced synthetic whistles that had the same frequency modulation but none of the voice features of known signature whistles.

> We conducted playback experiments to known individuals, testing their responses to synthetic signature whistles that resembled those of familiar related and unrelated individuals. We hypothesized that . . . animals would turn more often toward the speaker if they heard a whistle resembling that of a related individual. (8293)

The hypothesis—that individual dolphins would turn toward the synthetic signature whistles of relatives more often—proved correct, with the effect demonstrated in nine out of fourteen test subjects (8293). Additionally, by looking at how the dolphins responded to whistles similar to, but not the same as, the signature whistles of family members, the researchers found that "individual rather than kin discrimination" was taking place (8295). Signature whistles thus seem to serve as names that are recognized by others in the group.

Signature whistles are developed when infant dolphins copy heard whistles and modify the sounds slightly to produce their own signature whistles (8295). Thus, dolphins' vocal learning "allows increasing interindividual variability of signature whistles while maintaining potential group, population, or species features in the signal," leading to whistles that vary greatly over distances and in different populations of the same species of dolphin, as do human dialects (8295).

In a much earlier study on the language comprehension of bottlenose dolphins in captivity, two dolphins were taught to understand imperative sentences, one in an acoustic language and one in a visual language (Herman et al. 1984). The dolphins learned both languages; they could follow instructions in novel sentences and could comprehend sentences that expressed a new relationship between objects. Even in new contexts and locations, directions could be followed "at levels far above chance" (129). This sort of understanding is useful in training dolphins for rescue and in more abstract tasks. Dolphin researcher Karen Pryor, in a book she co-edited, *Dolphin Societies: Discoveries and Puzzles* (1991), recalls one dolphin that was trained in a free-swimming environment to look for objects in the water that were "anything man-made and bigger than a breadbox" (346). This dolphin "soon found engine blocks, a movie camera, quite a lot of fishing equipment, and a World War II airplane" (346) This anecdote perhaps shows that dolphins can come up with abstract categories of objects, and can work quite well with those categories to perform tasks.[4]

WHALES AND CULTURAL CHANGE

Though I've spent most of this chapter looking at dolphins, there is some fascinating research on whales' vocal abilities. A paper titled "Cultural Revolution in Whale Songs" details observations made of humpback whales off of the eastern Australian Coast in the Pacific Ocean (Noad et al., 2000). Male humpback whales are known to sing songs; the male humpbacks within a population all produce the same song, but the songs change through time, and all singers keep up with the changes, "implying a cultural transmission and evolution"—the same thing that happens with some bird songs (537). And different populations of humpbacks, especially those separated by great distances, produce different, "apparently unrelated" songs (537). In this oft-cited observational study, researchers were lucky enough to catch a song as it changed, after two new members were introduced into the group: two whales from the Indian Ocean, off the west coast of Australia (537). This observation happened serendipitously; researchers were merely recording the whale tunes when the two new group members were introduced, sparking the radical (and completely unexpected!) change of tune. No manipulation of context or experimental variable was introduced by researchers.

During a north–south migration in 1995–1998, researchers recorded 1,057 hours of songs using hydrophones attached to buoys and boats. In 1995 and 1996, researchers observed the song pattern change slightly—a normal and expected evolution. But the song changed radically in the years from 1996 to 1997, after the introduction of two new male singers to the group of 82 whales:

> In 1997, the new song became more common. Most of the 112 singers produced either the old or the new song, but three used an intermediate song containing themes from both types. By the end of the 1997 southward migration, almost all males had switched songs, and in 1998 only the new song was heard. (537)

There is typically very little interchange between the whale populations on Australia's east and west coasts, and the songs of the two populations have continued to evolve independently; this radical change of tune with the introduction of only two new group members is especially notable because "there are no examples of radical song replacement initiated by a small number of immigrant individuals in . . . any. . . species of song-bird," making this finding especially startling (537).

MEMORY AND COGNITION IN CETACEANS

In an overview study entitled "Cetaceans Have Complex Brains for Complex Cognition," a large team of researchers sum up the neurological findings on dolphins and whales, arguing against those who suggest that cetacean brain size is merely a by-product of ocean temperatures during cetacean evolution. Rather, these researchers suggest that an "integrated" view, combining research from cetacean neurology, behavioral studies, and evolutionary science, implies instead that cetaceans' large brains "evolved to support complex cognitive abilities" (Marino et al. 2007). Cetacean brains are larger than those of most other animals; when brain size is put into a ratio with body size, cetaceans' ratio is beaten only by that of human beings. This ratio-of-parts finding alone does not tell us anything about their cognition, of course, but the researchers marshal behavioral, anatomy, and evolutionary studies to suggest that cetaceans, like human beings and primates, are also capable of cognition.

One must look back 95 million years to find the common ancestor of cetaceans and primates, and cetacean brains have been on an evolutionary trajectory separate from those of other mammals (like hippopotamuses) for 55 million years (0966). For this reason, "there is no reason to expect that cetacean and primate prefrontal cortical analogs would be, in fact, located in the same region of the brain" (0966). But for both primates and cetaceans, the prefrontal regions are "consistent with high-level cognitive functions—such as attention, judgment, intuition, and social awareness" (0966). And the brains of cetaceans have "a large number of large V spindle neurons. . . . similar to those reported to be unique to humans and great apes" (0966). In the researchers' estimation of cetacean neurology, cetaceans' brains have "structural complexity that could support complex information processing, allowing for intelligent, rational behavior," and the researchers point to a wealth of behavioral data that supports this conclusion (0968). Experimental or laboratory studies have demonstrated that cetaceans can

- understand symbolic representations of things and events;
- understand how things work and how to manipulate them;
- understand the activities, identities, and behaviors of others;
- understand their own bodies and behaviors; and
- have accurate and robust memories (0968)

The researchers label these abilities, or types of knowledge, as declarative knowledge, procedural knowledge, social knowledge, self-knowledge, and memory, respectively (0968). Cetaceans display behaviors in all of these categories. Dolphins' ability to both abstract rules and to understand instructions, as well as their complex acoustic abilities, point to a robust intelligence. Indeed, "dolphins are the only mammal, other than humans, shown capable of extensive and rich vocal and behavioral mimicry" (0969). And we have already seen tool-use behaviors observed in dolphins and whales: recall the sponging and the beach hunting behaviors detailed in previous sections. Cetaceans also clearly display social knowledge.

According to these researchers, dolphins' imitative abilities are "one of the highest forms of social learning," one that goes beyond the abilities of non-human great apes (0969). And studies of wild dolphins have revealed the complexities and diversity of dolphin social groups and culture, including "impressive cultural learning of dialects, foraging sites, and foraging and feeding strategies" (0970). Cetaceans thus exhibit both culture, which the study's authors define as the transmission of learned behavior, and multiculturalism, or "groups with different cultures using the same habitat" (0970); for example, killer whales in the North Pacific display several social tiers, each with their own vocalizations, feeding techniques, and play-time activities Advanced social learning skills and cultural characteristics have been observationally documented in bottlenose dolphins, killer whales, sperm whales, and humpback whales—the cetaceans on which the most research has been conducted (0970). And dolphins display self-knowledge; they are able to recognize themselves in mirrors, and have thus demonstrated "a rare ability previously demonstrated in the great apes and humans, and, recently, in elephants" (0970). According to these researchers, dolphins clearly exhibit "sophisticated cognitive convergences with primates, including humans" (0971). This cognitive sophistication that dolphins share with humans is also supported by other integrative studies; for example, Marino (2002) argues that cetaceans and primates share important similarities of mind: "social behavior, artificial 'language' comprehension, and self-recognition ability" (21). Despite differences in environment and evolutionary history, it seems that cetaceans share the cognitive traits that are thought to be precursors to species' development of ideas about rights, community, and intelligence.

Though we have now traveled far afield from tool use *per se*, the research about cetacean intelligence, like the crow research I examine in the next chapter, can help us to understand what traits are important in defining technological behavior on our emerging map. It seems that, given the array of the techniques cetaceans use, their linguistic ability, and the abstraction of which at least some cetaceans are capable, these species may be capable of changes in techniques, innovation, and planning behavior, despite their dearth of physically manifested tool artifacts. Signature whistles and whale songs encapsulate information, and require know-how; they should therefore be considered, like manufactured artifacts, to be products of technological knowledge.

CONCLUSIONS

Cetaceans—mostly dolphins—have been observed carrying or using some external objects, usually for an undetermined use. In the volume *Animal Tool Behavior* (Schumaker et al. 2011), the authors categorize cetaceans' tool use into the following categories: throw, bait/entice, dig, prop/balance and climb, contain, and wipe (70–72). Humpback dolphins have been seen to *throw* seashells during play. *Bait/entice* behaviors include bottlenose dolphins' presenting of feathers and stones to researchers, and Amazon river dolphins' carrying and display of objects for undetermined purposes. The authors cite as *digging* behaviors the aforementioned sponging techniques of bottlenose dolphins in Shark Bay. *Prop/balance and climb* behaviors include a captive, ill bottlenose dolphin's use of a basketball on his rostrum; his caretakers thought he did this to reduce his effort in floating and surfacing to breathe. *Contain* behaviors include humpback whales' use of bubbles to help them catch prey (there are several hypotheses for how the bubble technique works). And, last, observed *wiping* behaviors include a bottlenose dolphin's use of a seagull feather and other items to wipe his tank window, seemingly imitating the way human divers cleaned his aquarium's surfaces. (He emitted sounds similar to those made by the air valve in a diver's suit while doing this.) These items, some of which are anecdotal, take up three pages of a book that categorizes and sorts tool use across many species—the sum total of current knowledge about cetaceans' use of external objects as tools. Clearly, given their conceptual and linguistic abilities, more observation is needed; however, cetaceans seem to be good candidates for users of technological knowledge, despite the small amount of observed evidence to date.

In the context of cetacean environments (which have poor visibility) and anatomy (their appendages are not particularly dexterous), the techniques used by cetaceans make a nice study for philosophers of technology and biology who are interested in the role played by environments, niches, language, and intelligence in the development of technology. Whether through the use of multiple techniques in hunting or foraging, through changes in vocalization or in signature tunes, or through displaying behaviors considered to be hallmarks of humanity (social behavior, language comprehension, and self-recognition), cetaceans' technological knowledge is an important addition to our growing map of technological behavior. Cetaceans clearly respond, use, make, and share.

There are several important takeaways from these cetacean case studies. First, the concept of *know-how* (useful, action-related knowledge about the material world and its transformation) offers a solid frame for thinking through these, and other, animal cases. Using techniques only briefly described here, dolphins work with the material world to reap its benefits (such as increased prey catching); they understand and carry out techniques in their environmental context. Although their material culture is limited both by their environment and its available materials and by their bodies, the role played by knowing-how in cetacean life is undeniable, and, in later chapters, the know-how displayed in these dolphin cases will be compared to the know-how of other animals.

Second, the dolphin cases can help to highlight and enrich discussions about what role society and culture play in the transmission of techniques and tools. The cetacean studies show that techniques can vary between and within groups, and they highlight the role of social learning. They also provide insights into the role played by planning and coordination in executing these various techniques.

Third, though it is not discussed in these terms in the literature, it seems plausible that dolphins can innovate. Given what we know about how the sponging technique was transmitted and employed, it makes sense to talk about dolphin hunting techniques in terms of innovation and culture. This "plausibility" move in my argument will be used again in later chapters.

What I hope to emphasize, in this chapter and in my discussion of the studies included here, is how cetacean cases help us broaden our understanding of technology and tool use. Cetaceans, too, like humans and like primates, should be seen as makers of tools and as users of technological knowledge. Insofar as these cases offer a glimpse at know-how and its making and shaping, the social transmission of techniques, and an ability to innovate and change, dolphin cases should be taken as relevant to discussions about technological behaviors and knowledge more broadly. Though cetaceans do not actively make large-scale things, they do use items in their environment in new and creative ways and arrange themselves socially to deploy new techniques. There is no good reason to think that intelligence, planning, and shaping are qualities limited to humans, or even to primates. Indeed, what we know about dolphins starts to unravel the human clause. The human clause cannot simply be stretched a bit to gather in primates; it must be rather radically revised or discarded altogether.

One way of folding the cetacean cases into the human clause would be to simply add another set of species to the human-primate clause: to say that technology is a method by which *minds like ours* negotiate the world through material artifacts. The literature on cetacean brains indicates that they are "like us"—more like us in brain-to-body ratio, at least, than other non-primate species are. The literature also indicates that cetaceans process information "like us"—they can form abstract concepts. Both of these points seem to support a simple expansion or revision to the human clause that we have found in definitions of technology. However, in the next two chapters, we will look at cases that are *less* "like us," and will discover that no human clause, however expanded, can contain the array and variety of tool use and technological behaviors that have been observed among members of some non-primate and non-cetacean species.

NOTES

1. Language requires form and manipulation and sometimes planning and technique and culture, though certainly of a less material kind than tool making would indicate.
2. My claim here—that language abilities are technological—might seem somewhat controversial, but there are defenders of this position. I will address concerns about this aspect in my concluding chapters.
3. There is actually now research on this topic that suggests that male chimps have pant-hoots that identify different groups, using a type of learned vocal matching.
4. Pryor also speculates about dolphins becoming the next animal to be domesticated by humans because of their usefulness in tasks like these, as well as their enjoyment of the sorts of tasks that trainers teach them to do. They may gain a role similar to other working animals, given tasks that "require an animal to give up the extensive social contact of a large wild group," but "gain the benefit of interesting things to do" (Pryor and Norris, eds., 1991, page 346).

SIX

Birds

The most recent revised and updated version of the book *Animal Tool Behavior* (Beck et al. 2011) catalogs tool behaviors by different species. The section on birds consists largely of descriptions of birds using tools, with fifteen pages on birds' tool use and with six of those pages about tool manufacture on studies of the New Caledonian crow. This chapter also focuses on the prolific tool use and manufacture of corvids: primarily New Caledonian (NC) crows, with some attention to the behaviors of rooks and scrub-jays.

The information I offer in this chapter, and the level of detail that I offer, are important to appreciating the complexity and detail of tool-related behaviors and technological knowledge in animals. Too often, animal tool use and tool making are dismissed as less meaningful and less cognitively complex than humans' similar activities, but it is very difficult to dismiss these cases when they are examined in detail. And these bird cases are especially important, because they are examples of tool use by animals that are very distant from human beings. Primates are seen as practically proto-human, and cetaceans are seen as "like us" because of their brain-body ratios and communication skills; bird cases, which are seen as more purely animal, pose more difficult challenges to maintaining a human clause in our definitions of technology and tool use.

Much of the recent research on birds reveals a sophistication in tool use that even chimpanzees cannot match. Birds certainly have the ability to manipulate and build objects—one only needs to look at a nest to realize that. Their beaks can be adept at shaping and sorting. But some bird species are doing things much more surprising and complicated than the simple object manipulation required for nest-building.

In this chapter, we turn to research on New Caledonian (NC) crows, rooks, and scrub-jays that is especially important for anthropologists and philosophers of technology. Many of the definitional features of technology can be found within the tool making and techniques used by these types of birds: NC crows manufacture distinct tool types; rooks perform intelligently in tests of calculation; and scrub-jays have shown an ability to plan (and to lie). We've already seen that the human clause cannot rule out chimpanzees, and we've seen that a modified hominid clause does not accommodate dolphins very well. Dolphins, of course, are adept at manipulation, at least of the linguistic sort, and this, along with their observed cognitive and behavioral similarities to primates, seems to align them with hominid tool users. But even a hominid-cetacean clause proves too limited in the face of new research on birds. *Whatever the hallmark of humanity is, it is not technological capabilities alone; the sorts of things humans do and the tool behaviors that humans display are not unique to humans alone; they are shared with other species in the animal kingdom.* Differences turn out to be a matter of degree, rather than a matter of kind;[1] even species whose brains cannot be well-analogized to those of human beings are capable of complex tool use, planning, social learning, and language.

NEW CALEDONIAN CROWS USE AND MAKE TOOLS

For the past twenty years, crows on the island of New Caledonia, in the Loyalty Islands in the Pacific Ocean, have been noted for their impressive tool use; observational research demonstrates the complexity, flexibility, and diversity of their tool-related—even technological—behaviors. Gavin R. Hunt's 1996 article in *Nature* marked the beginning of the reporting on the NC crow's tool use, describing the crows' manufacture of two types of hook tools that they use to capture prey: a hooked twig and a stepped-cut barbed pandanus leaf. The manufacture of these two types of tools shows "a high degree of standardization, distinctly discrete tool types with definite imposition of a form in tool shaping, and the use of hooks" (Hunt 1996, 249). These features of tool manufacture "only first appeared in the stone and bone tool-using cultures of early humans after the Lower Palaeolithic" (249). These crows have also been observed to manipulate twigs, drop nuts onto rocks to get seeds, and carry their tools around for use (249).

The hooked-twig tools that Hunt describes were made by the NC crows from living twigs, which the crows stripped of leaves and bark and "carved" into a shape with a hook on the end; the stepped-cut tools were cut from pandanus leaf edges, which were shaped by the crows into a taper, remaining sturdy and sharp with "their uncut edges . . . always faced upwards from the narrow ends" (250). With these tools, the crows use "slow and careful movements . . . to obtain prey from bases of leaves and holes where it was possible that crows could see their prey" (250). Hunt was able to collect tools from three forest sites, demonstrating that the tools were standardized; the diversity of tools found suggests the use of tool kits, and he suggests that the birds understand the tools' functionality (as indicated by the way they use the hooks) (250–251). Research performed after this foundational study points to an even greater complexity in crows' tool use behaviors; a much more recent study showed crows using three tools in succession (a tool set) to solve a problem to get food, and observed the crows crafting tools out of materials that they had not previously encountered (Morelle 2010). In fact, tool manufacture by NC crows in laboratory settings has been startling.

This Tool Use Is Learned

In tests in captivity, four naïve juveniles in captivity "developed the ability to use twig tools" without ever coming into contact with adult crows (Kenward et al. 2005, 121). These crows, when left with pandanus (similar to palm) leaves, were able to tear and cut them into shapes, some of which would have made decent tools, according to researchers (121). However, though they seemed able to cut and shape twigs and leaves, and although one even used a made tool to get to food, it seems that "social input . . . may be important in transmitting specific techniques and toolshapes," an idea bolstered by "the close attention . . . juveniles paid to demonstrations of tool use by their human foster parents" (121). There exists regional variation in the tool shapes of NC crows, who live in social groups, and researchers speculate that these variations may be "the result of cumulative *cultural* evolution" (121, my emphasis). While NC crows may have some innate capacity for the development of tools (as seen by the juveniles' leaf-shaping behaviors in captivity), it seems that the forms of tools produced and the materials used can vary significantly, and research suggests that social-cultural components may account for the variation. In other words, these crows may be inclined toward tool use, but their choice of materials and form may be greatly influenced by social factors.

Researchers Gavin R. Hunt and Russell D. Gray, both long-established researchers into NC crow behaviors, observed an adult crow and its dependent offspring manufacture ten tools (they caught some of the juvenile's manufacturing on video) and watched the crows use the tools (Hunt and Gray 2004, S88). For all of the ten tool-manufacturing sessions observed, the manufacturing process followed four basic steps:

> They (i) selected a fork formed by, usually, two twigs; (ii) broke off one twig just about the junction (side twig), then discarded it; (iii) broke off the remaining twig just below the junction (tool twig); and (iv) carried out fine sculpting of the hook on the tool twig with the bill, in between

> removing the compound leaves . . . The sculpting removed small pieces of wood form the hook, which refined and sharpened it. (S88)

Researchers timed the fourth step, which took, on average, 68 seconds (S89). The same techniques were used by the two birds: "a 'snapping' technique to remove side twigs and tool twigs," performed close to the bases of the twigs at the junctions, formed hooks near the tool ends (S89). However, the juvenile crow seemed to have less experience in making hooked-twig tools.

> Three differences . . . suggested that the juvenile was less experienced. . . . First, only the juvenile repeatedly worked on the hook after first use of the tool. This behavior may have been a reaction to limited success at extracting food. . . . Second, after picking up tools from a horizontal position the juvenile sometimes used the non-hooked ends as the working ends. . . . The parent never attempted to extract food using the non-hooked end of a tool. Last, the juvenile crow had noticeably more difficulty in snapping off the tool twig than did its parent. (S89)

These differences, given the respective ages and relation of the two crows, suggest that the exact manufacturing process is something passed from adults to juveniles (and that juveniles may steadily learn how to craft a particular tool from their parents, though examination of this is outside the scope of this crow study's discussion).

The Tools Are Crafted

Crafting, according to scientists, has three components: "(i) selection of an appropriate section or piece of material, (ii) preparatory trimming and (iii) fine, three dimensional sculpting" (S89), a process that is typical of both crow crafting and "early human tool manufacture" (Hunt and Gray 2004, S88). Although, according to the researchers, tool manufacture by animals is common among woodpecker finches and chimps, the "fine sculpting" seen among NC crows had been previously undocumented in any other species besides human beings (S88). The crows' manufacture of pandanus-leaf tools, which is complex, may not be considered crafting or sculpting because it is two-dimensional and "no preparatory trimming is carried out" (S88); however, their manufacture of hooked-twig tools, which are molded in three dimensions, does seem to be a form of fine sculpting that was previously thought to be achieved only by humans.

Hunt and Gray's discussion of the crows' hooked-twig manufacturing refers back to human manufacturing capabilities:

> New Caledonian crows appear to have a rudimentary technology analogous to that of early humans. This rudimentary technology includes the cognitively demanding task of crafting tools. . . . This routine of complex manipulations [the four-step process described above] is con-

> sistent with the possibility that the crows' goal was the manufacture of a hooked tool. . . . The tool manufacture of NC crows has four features previously thought to be unique to hominids: a high degree of standardization, the use of hooks, "handedness" and cumulative changes in tool design. (2004, S89–90)

The authors also point to laboratory observations of captive crows bending wire to get food—an observation that suggests that "crows may have at least a rudimentary grasp of the physical properties of objects or 'folk physics'" (S90). This term "folk physics" emphasizes again the similarities between NC crow tool manufacturing and that of humans, and we will revisit this term shortly in more depth. NC crows in the wild have been found to work in "parallel tool industries" that were once thought to be unique to humans, a legacy of early human hunter-gatherer societies (Hunt and Gray 2007). In observations of twelve crows on Maré Island, the two researchers documented individual crows' strong preferences for particular tools and their use (173). Eight of these crows exclusively used one type of tool—either a pandanus or stick tool—and all of the crows showed a preference in use for one or the other tool (seven preferred stick tools and five preferred the pandanus tools) (173–174). In the crow societies, tool preference was not gendered, although in early human societies it did tend to be so (likely due to "social conformity" reinforcing the division of labor by sex) (174, referencing Bird 1999). The crows' documented specializations in tool use "could either be a consequence of genetic differences or different, vertically inherited (parent-to-offspring) social traditions" (174). Though captive juvenile crows have been shown to have a disposition for the employment of tools, the social organization of these crows in the wild, and especially the amount of time adults spend with their dependent juveniles, increase the likelihood that tool preferences develop from specific tool-skills that are transmitted from parent to juvenile (174).

Do They Understand What They Are Doing?

"Folk physics" refers to accounts of the world that reflect common wisdom or a rudimentary understanding and prediction about events in the physical world. In a study conducted by Alex Weir, Jackie Chappell, and Alex Kacelnik, a female NC crow in captivity was able to effectively shape wire to poke down a tube to get food, demonstrating an understanding of "folk physics." The researchers presented two crows—one male and one female—with a tube that had food at the bottom and an unbent piece of wire (Weir et al. 2002, 981). These two crows had seen similar tube apparatuses in the past, but their only experiences with pliant materials had been a year previous, when they had encountered pipe cleaners. The wire was entirely new to them (981). In ten trials, the female crow successfully bent the wire and got the food nine times; the male did not bend the wire (981). In all successful attempts, the food was retrieved within two minutes (981). The female crow used no technique found in normal crow life to bend the wire, and the technique she used "would be unlikely to be effective with natural materials" (981). This crow, who lacked previous experience with pliant materials or wire and had no model to imitate, carried out a "purposeful modification . . . without extensive prior experience"—the first time this behavior has been observed in non-human animals (981). The researchers note that "in experiments by Povinelli, chimpanzees (*Pan troglodytes*) repeatedly failed to unbend piping and insert it through a hole to obtain an apple, unless they received explicit coaching. Further experiments have shown a similar lack of deliberate, specific tool modifications in primates" (981).[2]

The observations made in the wire-bending experiment, say the researchers, especially "in a species so distantly related to humans and lacking symbolic language," point to important questions about "the kinds of understanding of 'folk physics' and causality available to nonhumans, the conditions for these abilities to evolve, and the associated neural adaptations" (981). NC crows' understanding of folk physics was further documented in a 2004 study by Chappell and Kacelnik. In a previous, separate study by the same authors, crows had been shown to be able to select tools of appropriate lengths for various tasks (Chappell and Kacelnik 2002). The same crows were then tested for the ability to select tools of appropriate diameter for various tasks. Two subsequent experiments tested whether NC crows were capable of determining the suitability of a tool for a given task (Chappell and Kacelnik 2004). In the first experiment, the female crow[3] was presented with three pre-made tools of different diameters and food in a tube with a hole; of the three tools presented, two were tied in a bundle; for the tools to be used, the bundle had to be untied (122). One tool was too thick (the "thick" tool), one tool was just thin enough (the "medium" tool), and one tool was "thin enough to serve in all conditions" (the "thin" tool) (121–122). The female was found to prefer the thin tool in almost all cases, even when it was bundled and even in post-testing phases; when all the rods were unbundled, the female chose the thin rod in 24 out of 24 trials (123). The female crow untied the bundle only when "the thin rod was not freely available" (126). Researchers postulate that ergonomics may have partially determined this choice, as the thinner rod is necessarily lighter.

In the second experiment, the female and male crows were tested together. The food tube apparatus remained the same, but instead of the sticks, the crows were left with a "fresh, bushy branch of oak" (124). Each branch had many twigs with diameters thin enough to enter the hole in the tube. Trials went on for 30 minutes or until one of the crows got the food; only one trial was ended due to time elapsing, and in all other trials the crows retrieved the food within 20 minutes (124).

The crows both made and used their tools in these tests. The female obtained the food in 16 of 17 trials, and the male obtained food in all 13 trials in which he participated (124). The researchers describe the process:

> In each trial, the crows approached the tube and looked at it, either from the nearest perch . . . or they landed on the table next to the tube. . . . We are therefore reasonably confident that they had the opportunity to assess the size of the hole before making a tool. They then flew to the branch, snipped the leaves off large areas of twigs, and then finally removed a twig. . . . In only two trials did one of the crows (the female in both cases) modify a tool after she had attempted to insert it, by removing projections that prevented insertion of the tool in the hole. (124–125)

The crows' manufacture of "such a well-fitting tool" is particularly notable, say the researchers, especially when the "uneven nature of oak twigs and sticks" is considered (126.) Researchers observed the crows examining the situation, assessing the specifications for the task, and then creating the appropriate instrument. Furthermore, when researchers changed the size of the tube's hole, making it thinner, the birds produced in response tools that were thinner, demonstrating "some level of understanding of the requirements of the task" (126). The researchers explain that "crows certainly have the capacity to adjust the specifications of the tools they make to suit the task at hand. Furthermore, *they appear to understand the function of tools,* and (at least one of the captive crows) can make appropriate tools using novel manufacturing techniques and materials" (126, my emphasis added because wow). I rely greatly on quotations in this section because the study's authors describe crow tool use and manufacture using language that is very suggestive about crows' technological understanding and behaviors—language that, given my particular philosophical research agenda, might sound biased in paraphrase. And while the researchers certainly have incentive to present novel, exciting bird behavior, they do have direct experience with the birds, which gives them authority to speak about what the birds understand and are capable of.

And the language of the crow studies researchers—"crafting," "adjusting to specifications," "understanding the requirements of a task," "appropriate tools," etc.—sound more appropriate to engineering philosophy than it does to animal studies. In fact, at times, these crow researchers sound as though they are reporting on engineer behaviors: these crows' "appropriate technology" is both fit to task and flexible to environmental/experimental requirements.

NC Crow Case: Behaviors Beyond the Capabilities of Chimpanzees

In what I find, from a philosophy of technology standpoint, to be the most provocative study of crows in captivity, Weir and Kacelnik report that Betty the crow "creatively re-designs tools by bending or unbending" (2006). In one study, researchers hoped to further establish the NC crows' sophisticated understanding of physical tasks—one unrivaled by any other non-human animal. They explain that

> the act of making functional artefacts is often thought to be especially revealing about cognitive processes, because it may require reference to both the representation of the problem and the expected future use of the artefact. However, this assumption is not always valid: the artefact maker might be simply following action rules acquired by the species through natural selection. . . . In other cases, individuals may learn through trial-and-error, or by observing others, what sequences of actions modify the artefact effectively, but again with no cognitive representation of the problem or planning the future use of the instrument. (Weir and Kacelnik 2006, 317)

The problem with animal behavior research lies in the difficulty of establishing what *exactly* is indicated by a behavior, and the same difficulty holds in examining animal tool use. The chimpanzee studies seem to indicate that trial-and-error and observation play an important part in chimp tool making; similarly, dolphin sponging and other learned foraging techniques may be communicated via observation, and imitation. But how can we know if something more than imitation or trial-and-error[4]—or more than a simple stimulus-response written into the animal's genetic make up—is at work? Is this ability to go a step beyond—to plan, to represent in our minds—the thing that sets humans apart, that makes our tools technological? Is this the dividing line?

While the studies we have examined so far do not help us answer the first question, the studies of dolphin naming and calling seem to indicate that cetaceans do make mental representations, and social learning almost certainly plays a role in whale calls. The chimpanzee tool studies seem to indicate that trial-and-error and observational learning are extremely important. Language studies of great apes in captivity perhaps best demonstrate the cognitive processes of which chimps are capable, although ability in captivity doesn't always translate to the emergence of the behavior in the wild.

Studies of NC crows, however, can perhaps begin to answer the first question. These case studies are of animals much further from and less easily analogized to the human species; these are also studies of animals whose linguistic abilities are not often studied (although the calls and whistles of other birds are studied, crows' calls rarely are). While animal tool use and tool making can demonstrate cognitive processes and planning, we must carefully observe how the process plays out; flexibility in materials and methods, the use of novel techniques, and new solutions for new problems all seem to indicate something more than stimulus-response behavior or trial-and-error learning. These things seem to indicate creative cognitive processes in the face of new situations. In the words of Weir and Kacelnik (2006), "the more innovative the actions and the more specific, deliberate, and unusual the modification of the raw material, the more acceptable it becomes to hypothesise that the agent's behaviour is controlled by cognitive representation of a definite goal and the means of reaching it" (318).

While no one test can conclusively show that a species has the same types of cognitive abilities as human beings, compiled data on several species seem to point to flexible, creative, and inventive tool-making behaviors.

One Research Subject In Depth: Betty

One particular crow, Betty, was the subject of a series of trials involving bending wire to use as a hook, and a further set of three experiments with a novel material—aluminum strips—to see if Betty understood the physical tasks (Weir and Kacelnik 2006.). This study did not only test the crow's causal reasoning, but also her understanding and flexibility. In earlier studies, Betty "spontaneously and repeatedly" bent wire, making hook shapes to get a tiny bucket of food out of a tube. To her handlers, Betty's actions seemed to meet three important criteria: the actions seemed deliberate (hooks made and used immediately for their task); the actions seemed specific to the situation (she had never made this type of tool before); and the actions seemed novel (she was acting on/with something she had not encountered before in the lab or as part of any native habitat) (Weir et al. 2002; Weir and Kacelnik 2006, 318). The authors admit that no single study can clearly establish these capacities (Weir and Kacelnik 2006, 318). They suggest that, in studying animal understanding, researchers should attend to how long animals take to get from a problem to a novel solution. They suspect that:

> a slow and gradual increase in proficiency would indicate that the subject relies on within-task trial-and-error learning. In contrast, immediate or step-wise acquisition would suggest that, at a minimum, the subject generalises from concepts formed during earlier experience in related tasks. . . . [T]he greater the understanding of the problem (namely, the degree to which abstracted general principles play a role), the greater should be the flexibility shown to produce novel transformations in line with new demands. (ibid., 318)

In the study in which Betty made hooks from various materials, the researchers used three modifications of the problem to examine whether Betty could come up with a novel solution and how her problem-solving process would unfold. Experiment 1, which had the same setup as her 2002 trial, required Betty to bend a novel material to make a hook to retrieve a treat. Instead of wire, Betty was given a very different material:

a thin strip of extremely pliable aluminum (321). There was no prior coaching on the material, and there was no pre-training on the setup, since Betty had already encountered it. For some of the trials, Betty was allowed to enter the testing area freely; sometimes she came prepared with her own tool—a twig or feather (323). According to researchers, "Betty adapted quickly to the new material." She used the aluminum strips by the third trial; obtained the bucket using the strips in 25 of 34 trials; obtained the bucket with her own tools in four of the trials; and lost the aluminum strip, either in the tube or behind the setup, in the remaining five trials (323). Betty's initial trials, which were less successful, took longer than the later trials. The researchers describe the overall outcome and what it suggests:

> Firstly, she learned very quickly how to effectively modify the tool, even though she had to use completely different techniques from those used with wire. In fact, from the sixth trial onwards, she only once spent more than 5 s crafting the tool. In addition, the "hook-ness" of her tools rapidly improved and became more regular . . . , although even towards the end of the experiment there were occasional malformed ones. This is despite the difficulty of modifying this kind of material with a beak as her only manipulative appendage, and the fact that the modification techniques she used are unlike any known actions used by wild crows, or by her in other circumstances. (326–327)

One would expect that a simple trial-and-error learning process would have taken more time and that Betty's previous work in bending wire would have led her astray in this task. However, Betty's performance in the first experiment suggests that there is more going on in Betty's understanding than simple trial and error.

Researchers set up the Experiment 2 with Betty such that "the successful actions from the previous problem would lead to failure" (327). Betty was given aluminum strips, bent at both ends, but Betty now had to retrieve her snacks from a narrow hole. Experiment 2, which was conducted at the tail end of Experiment 1, is modeled on some chimpanzee experiments done in captivity (327). Betty was familiar with this narrow-hole set up from prior experiments with other researchers, and the rewards for Experiment 2 were the same as those in Experiment 1 (328). To obtain the treat, a sufficiently narrow tool needs to be inserted into a hole, pushing the treat cup far enough along to fall down a bend in the tube and onto the ground. The now-bent aluminum strips that Betty was given, which were already shaped into hooks on both ends, could not fit into the hole while bent into hook form (328).

Before testing, researchers made sure Betty was given practice trials with this new setup. In the practice trials, for which she was given straight tools, she was able to obtain the food. In the three official experiment trials, Betty quickly retrieved the food in every case, although in two instances not as the researchers intended; in trial 1, Betty pecked hard at the hole and the food cup was jostled enough to fall—a successful, if unexpected, method of getting the food. In trials 2 and 3, she used the aluminum strip. She used the strip without modifying it in trial 2; she stuck it up the vertical end of the pipe and used the hook to bring the treat cup down (328). In trial 3, researchers were wising up to Betty's new methods, and they modified the vertical tube so that a repeat of her technique in trial 2 would have been impossible. In trial 3, Betty "performed the task as intended" by the researchers, squeezing both ends of the tool to flatten it, enabling her to poke through the hole, "thereby solving the problem of 'spontaneously modifying the tool to allow it to fit through the hole'" (328). On this third trial, she initially tried to use the wrong end of the tool, but quickly turned it around. This "could be interpreted as Betty instantly 'understanding' what was required," since she did not continue to try it with the wrong end (328).

The second experiment used a setup similar to previous chimpanzee studies, in which all seven chimpanzee participants "showed a strong preference for attempting to insert the 'impossible' ends of tools, and very rarely turned the tools around" (328, and referencing Povinelli et al. 2000). In 56 chimp trials, there were only three successes. The researchers who worked with Betty conclude that, "in this context, Betty's response of *turning the tool around almost instantly is impressive, even if not equivalent to human-like understanding*" (328, my emphasis). In Experiment 3, researchers were looking for something more conclusive. They presented Betty with an instrument that was more difficult to squeeze and modify; the aluminum strips were now bent in a U-shape that would need to be straightened out, and Betty had to make the instrument longer instead of narrower in order to successfully retrieve her treat (329). No training on the apparatus was given, since Betty had encountered it before, and Betty was given four trials (329).[5] She successfully obtained her treat in three of four trials. In the first trial, she "managed to reach and retrieve" her reward "by inserting her head and neck into the entrance of the tube" with the ends of the tool squeezed together to make a straight tool; the apparatus was subsequently modified to prevent Betty from using her head and neck to get the material in the tube. On trial 2, Betty tried the unmodified tool for 1.5 minutes without success. On trials 3 and 4, Betty modified the aluminum strip to get the reward.

> Both trials involved a similar modification technique, which occurred several minutes into the trial: in the middle of a bout of probing in the tube, she raised her head and beak (still holding one end of the tool) in a distinctive and unusual manner, causing the shaft of the tool to bend backwards against the lip of the tube. (329)

While researchers cannot prove that this odd behavior was purposeful, it had never been witnessed on other occasions with different materials (329). Perhaps what is most convincing about these experiments is how Betty adapted her tool use for each problem. In Experiment 3, "in no trial did she modify the strip by bending or twisting it, which were the actions associated with success in Experiment 1" (329). This suggests to the researchers that Betty recognized that those actions wouldn't work in Experiment 3—and she realized this without recourse to trial and error! Betty used strategies that suited the problems with which she was presented, even when prior training could have interfered with her possible strategies, and she showed the researchers problems with the apparatus on several occasions, coming up with unanticipated ways of getting to her treats.

In their discussion of their experiments with Betty, the study's authors reflect on how Betty's understanding may relate to human understanding:

> While her *innovative behavior* cannot be accounted for purely by reinforcement for specific actions, it is not yet justified to assume that she possesses a full, human-like understanding of each task and that she uses it to plan and direct her behavior (although whether the full understanding that humans presumably have of the task would reveal itself by perfect first-trial performance is unclear, since humans often make mistakes despite such understanding).[6] (331, my emphasis)

The researchers here emphasize three points. The first is that just because a subject performs a task in a way not foreseen by humans, it does not necessarily follow that the subject does not understand the task requirements (331). In some studies, for example, humans have performed tasks such that observers could believe they do not understand gravity; the researchers are thus quick to note that "it is critical to test how humans perform on tasks that they do understand before interpreting a non-human animal's failure as evidence for lack of understanding" (331). I would add to this point that problem-solving strategies that work for non-human animals probably differ from those that work for adult humans (just as a toddler's understanding is different from an adult human's, and as a juvenile dolphin's differs from that of an adult dolphin). Because of environmental and habitat differences, different animals are differently attuned to the world. This may explain why it is so difficult to interpret whale and dolphin studies: their environment, spatial processing, and evolutionary history are radically different from ours.

Their second point is that "some progress can be made by comparing behavior of members of different species in comparable tasks," and that this can be achieved by designing studies to mimic other experiments. For example, the crow researchers who worked with Betty often make reference to chimpanzee studies, and they even designed some of Betty's tasks to compare directly with those reported in the chimpanzee literature, enabling direct comparison. Betty, of course, performed much better than the chimpanzees, "who are often considered to be the most intelligent non-humans" (331). Betty's authors do note, however, that NC crows develop tool use even when raised in isolation, and that the wild birds use tools widely in their natural environments; it is possible, then, that "these birds might . . . have specific cognitive adaptations that make them particularly good at learning and possibly reasoning about tasks involving physical interactions between solid objects, but perhaps not extraordinary at other equally difficult logical tasks" (331–32).[7]

The third note the authors make is that understanding is usually seen as all or nothing, but this is a false dichotomy; it is more of a spectrum, for we can see many intermediate levels of understanding (332). The authors suggest that when thinking about levels of physical understanding or "folk physics," it may be useful to think about "the minimum specifications that need to be incorporated into a robot to achieve a similar level of generalisation and creativity" (332). The authors also point out that our perceptions of understanding may be affected by, for example, sensory and perceptual differences between species, which might account for disparities in data on animals' causal understanding. Much other exciting research has also been published on NC crows and tool use in the past fifteen years, from studies of "parallel tool industries" among the crows like those seen in human hunter-gatherer societies (Hunt and Gray 2007) to "human-like, population-level specialization" in the manufacturing of pandanus leaf tools (Hunt 2000), to studies of handedness and social learning (Kenward et al. 2006), to decision strategies for tool use (Hunt et al. 2006) and metatool use (Taylor et al. 2007; Taylor et al. 2010). There is even a study that directly compares crows' and humans' tool selection for a given task (Silva and Silva 2010). I touch here on just a few more crow studies before turning to other bird species for the remainder of the chapter.

Betty passed away in 2005, but observations about her set an agenda for research into bird tool use and problem solving. Most recently, researchers have studied NC crows in the wild, observing the crows bending and fashioning tools in their native habitats, and they have studied ten subjects that use these techniques. Although Rutz et al. argue that these bending and shaping behaviors might be part of the NC crow repertoire and try to explain Betty's activities as "pre-existing tool-manipulation routines" (2016), the rook cases we will look at in the end of this chapter lead us away from this conclusion.

And They Use Tools to Make Tools

Metatool use, or the use of one tool on another, is considered by some to be the first step from tool use to technology; human development of metatool use is considered a "major innovation" for human evolution, one of the things that propelled us to our current technological world.

Metatool use presents "three distinct cognitive challenges" (Taylor et al. 2007, 1504). For an individual to be engaged in metatool use, they " must recognize that tools can be used on nonfood objects," they "must initially inhibit a direct response toward the main goal of obtaining food, a reaction that both children and primates find difficult to suppress," and they "must be capable of hierarchically organized behavior" (1504).

In addition to the ability to organize behavior hierarchically, the use of metatools requires *flexibility* of hierarchical thinking—an agent must be able to include a "newly innovative behavior (tool → tool) with established behaviors as a subgoal" (1504). This process "has been suggested to follow a recursive pattern and to require cognitive processing similar to language production" (1504). In one study, all seven observed NC crows "spontaneously produced the correct behavioral sequence" in a trial on metatool use (Taylor et al. 2007). The experiment was set up so that crows had to use one tool to get to another to extract food, and the researchers included some irrelevant tools to check for trial-and-error learning. The crows in this study did not "randomly probe the toolboxes" with unnecessary tools, and the researchers suspect that analogical reasoning played an important role in the crows' rapid solution to the task (Taylor et al. 2007, 1506).

But it is difficult to distinguish associative learning from causal reasoning. A subsequent study in 2009 on causal and analogical reasoning in NC crows opens with a philosophical point about how causation can be identified: "David Hume (1711–1776) famously used the example of one billiard ball rolling into another to illustrate his argument that causal relations cannot be explicitly perceived. Instead, causal relations must be inferred from sensory information" (Taylor, Hunt, Medina, and Gray 2009, 247). These crow researchers, like other animal researchers, struggle to tell from observed behaviors whether particular actions are the products of associative learning—the simple pairing of a stimulus and a response—or from causal reasoning, which entails understanding. In this crow study, the researchers employed a trap-tube and a trap-table, two types of apparatus commonly used in ape studies of causal versus associative learning. The two devices, which look quite different, operate on the same causal principles; for a successful trial, the subject must move the treat in one direction (but not too far) and then work on it from another aperture to avoid losing the meat in a trap (a hole in the bottom of the setup). They found that crows were able to extend the causal relations required to solve the trap-tube to the very different set up of the trap-table (251). Inhibitory control presented a problem for some of the crows tested, who were unable "to stop pulling meat towards itself and instead walk around the apparatus" to probe into it from that end (251). This inhibitory control caused failure in the tube task for three of the six crows tested, but this problem is one that is present "in problem solving for both children and non-human animals" more generally (251). One of the birds (named "Español") "hesitated before pulling the meat into the trap" (pulling the meat into the trap means the subject did not reason out how to get the meat out, and therefore does not get the treat) (251). This hesitation "suggests that Español may have learnt when he would fail, [but] . . . was unable to inhibit pulling the meat towards himself despite the presence of the trap" (251). The researchers carefully rule out as implausible non-causal explanations (chance, visual and tactile generalization, and prior dispositions) for the crows' behavior (251–252). The most plausible account of this study's findings, say the researchers, is that "the crows transferred knowledge of the causal relations between the hole and the reward to the perceptually distinct trap-table problem" (252).

In a follow-up study of reasoning in NC crows, crows had to "infer the presence of a hidden causal agent" (Taylor, Miller, and Gray 2012). I see this study as the mirror image of the 2009 study just described: metatool use means that the subjects successfully reasoned causation *forward* through time (one thing can cause another can cause another), and this study shows the subjects reasoning *backward* through what happened (figuring out how one observed thing was caused by some previous factor). In this 2012 paper, which I will not describe in detail because tool use is not featured prominently in the trials,[8] the results suggest that these animal behaviors may be underpinned by complex cognition.

> The ability to make inferences about why an inanimate object is moving would be highly adaptive in many ecological situations. . . . It is, therefore, possible that the ability to reason about hidden causal agents is far more widespread in the animal kingdom than has been thought previously. (Taylor et al. 2012)

As the authors indicate, we should perhaps expect causal reasoning to be *more* common among other species because of the survival benefit conferred by the ability to infer from movement the presence of predators and prey.

Researchers have also sought to understand NC crows' innovation and problem-solving skills. Crows were tested using a stone-dropping test that was developed for rooks (whose responses to this test are described in the next section); the problem involves dropping stones off a platform into a vertical tube to collapse a platform, freeing a treat (von Bayern et al. 2009). Six birds were used in the study. As in the rook study, two crows were trained in how to nudge stones into the vertical tube; these crows were used as a control group. Four crows were trained by witnessing how to push down the platform with their beaks. In the trials, two of the four crows that were *not* trained in stone-nudging nonetheless picked up stones and dropped them into the tube, though they had never before encountered this procedure (1965). This, and the behaviors of the birds in relation to the apparatus, led the researchers to the conclusion that "the ability to reason about invisible forces has not yet been convincingly demonstrated in nonhuman animals, but these recent observations with corvids may encourage future experiments" (1965–1966).

This seems to suggest that these crows have some understanding of folk physics and gravity: half of the crows who had not been exposed to stones were able to solve the task, one that required some understanding of force being laid on the platform for the release of a treat. Although researchers suggest that the crows' "insight" into the problem is hard to account for, they concede that research on crows and other corvids could lead to conclusions about these birds' cognitive processes that might challenge our inherited perception that man is the only intelligent animal.

Innovation by New Caledonian Crows

"Innovation" is a word typically associated with only human beings, but there are at least two studies on NC crows that point to these crows' "high innovation rates in the wild" (Taylor et al. 2010; Hunt and Gray 2003). The first of these studies, published in 2003, points to crows' "cumulative technological evolution"—something that is likely the product of innovation along their evolutionary history (Hunt and Gray 2003, 867). This study included a survey of New Caledonia intended "to establish the geographical variation in the manufacture of these tools"; this study, which specifically targeted pandanus tool manufacturing, documented 5550 tools at 21 sites (867). The researchers found three very distinct types of pandanus tools: a wide tool, a narrow tool, and a stepped tool (the tool described in detail by the same researchers, quoted earlier, was a stepped tool; Hunt 1996). Hunt and Gray describe three characteristics of cumulative technological evolution: diversification in design, "cumulative changes to tool lineages," and "faithful transmission" of design through social processes (Hunt and Gray 2003, 867). According to the researchers, "when a new tool design is added to one or more existing, related designs," the tool designs are being diversified (867). The crows' three distinct types of tools, all made from a single material, are likely to be diversified versions of the same tool; it is unlikely that these tool types developed separately, and researchers suggest that they come from a common origin, especially since initial cuts made to the pandanus leaves were likely similar (872).

What the researchers did *not* find was poorly made stepped tools, which might have indicated individual learning on the part of the crows; the researchers suspect that the fashioning of leaf tools implies social transmission, rather than trial and error (872). In fact, it appears that two of the three designs "must have evolved from cumulative change(s) to earlier versions" because "each design results from a unique, non-recapitulating manufacture process" (872). Hunt and Gray thus suggest an evolutionary history for the tools' development and conclude that their findings "are the first indication that a non-human species may have evolved rudimentary cumulative technology," which is likely socially transmitted (872). NC crows' innovation behaviors were further tested to see whether they could apply known strategies to novel tasks and contexts. The experiment was set up to see if crows could apply an abstract rule, "out of reach objects can be accessed using a tool," in a completely novel context (Taylor et al. 2010, 1). This three-stage metatool problem required the crows to perform six steps to get the treat (1). Crows were divided into groups—a "component" group, made up of crows who had been exposed to the individual components of the test before, and an "innovation" group, who lacked relevant contextual and metatool-use background (2). The three crows in the component group all successfully solved the task on the first trial in the way anticipated by the researchers (2). In the innovation group, two crows (Sam and Caspar) solved the problem on the first trial—one without making any errors and "after inspecting the apparatus for 110 s," and the other after stopping to reinspect because of a dropped string that he needed to complete the task (3). The other two innovation-group crows (Maya and Djinn) were able to complete the task, one on her third trial and one on her fourth trial (3).

The component-group crows in this experiment demonstrated that "New Caledonian crows can spontaneously link up to six learned behaviors into a novel behavioral pattern" (4). The behavior of Sam (the innovation-group crow who solved the problem on his first attempt without error) demonstrated that these crows can also spontaneously link these behaviors while using two new behaviors in a novel context (4). Sam's performance matched that of the three crows in the component group, despite the new challenges involved (4). This is not a product of conditional reinforcement, both because of the complexity of the process and the use of metatools required. Other simple cognitive mechanisms are also ruled out by the performance of the innovation group; while the component group had the appropriate "behavioral repertoire" and "functional generalization," the innovation group lacked all relevant experience—the innovation-group crows had to come up with new behavioral patterns in a new context, and their ability to do so points to deeper cognitive abilities than had been previously suspected (4). Researchers also ruled out the possibility that their behavior could be explained by resurgence, but that does not account for the innovation group's "generated metatool use" (4). Researchers also ruled out the possibility that the crows had a propensity for the activities required in the task—particularly the string-pulling (an activity with which the innovation group had no experience). However, in other experiments, NC crows have rarely been observed to pull on a string without meat attached to it, so string-pulling seems to be outside their normal behavioral repertoire (5). The "spontaneous" innovation is so defined because of its "immediacy and lack of explicit training" (5). In summary, the researchers conclude that the observed "behavioral innovation, particularly the [crows'] use of behaviors in novel contexts, can be underpinned by cognitive mechanisms that are more complex than, but supplement, simple learning mechanisms" (6).

Note that the researchers here use terms here that are normally associated with human technology and learning: "cumulative technological evolution," "behavioral innovation," "causal reasoning." And these traits are not limited to NC crows; they have been confirmed in other birds as well, although these other species' native tool use is not as sophisticated as that of the crows.

ROOKS CAN MAKE SOPHISTICATED TOOLS, EVEN THOUGH THEY DO NOT IN THE WILD

Like crows, rooks are corvids, but rooks are not known to produce any tools in the wild. However, in laboratory studies, rooks have been found to be "capable of insightful problem solving related to sophisticated tool use," indicating that there are species capable of the sorts of *reasoning* necessary for tool use and production who *do not actually produce or use tools* in their native environments. In a 2009 study, authored by Christopher Bird and Nathan Emery, four rooks were put through a series of trials after demonstrating the ability to drop a rock down a vertical tube to collapse a platform containing a worm. These four birds first learned to perform the task by accidentally nudging a stone or by watching other birds solve the task (Bird and Emery 2009, 10370). One subject in the study, named Fry, "spontaneously picked up the stone and dropped it into the tube" after watching "her partner Cook successfully complete the task" (10370). These rooks' behaviors fit Benjamin Beck's famous definition of tool use: "the external employment of an unattached environmental object to alter more efficiently the form, position, or condition of another object, another organism, or the user itself when the user holds or carries the tool during or just before use and is responsible for the proper and effective orientation of the tool" (Beck 1980, as quoted in Bird and Emery 2009), and "present the opportunity to investigate insightful reasoning" in follow-up testing (Bird and Emery 2009, 10370). After performing this initial task, the rooks' other trials were performed in isolation in a testing room that could not be viewed from the aviary, although it was open to fly into (so that birds could come in and be observed in the experiment) (10370).

There were many problems to be solved in these rook tests: selection of appropriate stone size, ability to retrieve the appropriate tool, decisions about the use of sticks and which of two sticks to use, a spontaneous metatool use task, modification of a stick, and tests with hooks and the manufacture of hooks. Many of these experiments were similar to setups presented to NC crows; the rook researchers were familiar with the crow studies and actively used them as a comparison. The rooks performed astonishingly well for a species that does not use or make tools in the wild. These four rooks were able to choose stones capable of collapsing platforms with worms—in other words, they chose the stones based on their "functional relevance" to the tests (10371). They found appropriate stones in their aviary when appropriate stones were not provided in the testing room, picking out stones of the appropriate shape to fit into the vertical tube of the testing apparatus (10371). These rooks also used a light, long stick to push down the platform, demonstrating the use of a completely new tool and their ability to use "goal directed action" instead of conditioned response (10371). Given the choice of one nonfunctional and one functional tool (sticks and stones of varying weights and sizes), rooks correctly chose the functional tool "regardless of the tool type" (10372). Faced with a metatool task involving the use of "a large stone to access a small stone that could be used to release the inaccessible food," all four rooks solved the task on the first trial (10372). When tested on their ability to manipulate a stick tool, rooks were able to tear off the twigs on an elm stick, successfully modifying the tool to allow them to use it to access a treat (10373). Rooks tested on a New Caledonian crow task—using a hook tool to obtain a bucket of waxworm—performed successfully (10373). A second phase of hook-tool experimenting had the rooks presented with a choice of two hook tools, one functional hook and one that would not work. More often than not, the rooks in this test chose the functional tool, with three out of four rooks choosing correctly in the first trial; researchers claimed that the rooks "actively discriminated" between the two tools (10373). In the final phase of hook-testing, rooks were presented with straight wire that would need to be made into a hook to pull up a bucket with a treat from a vertical tube—an experiment done with Betty. The results were surprising, especially given that only one NC crow had been shown to perform this task: "all 4 rooks spontaneously manufactured a hook and used it successfully to extract the bucket, 3 of the 4 subjects achieving this on the first trial" (10374).

The tests with these four rooks show that rooks' tool-use abilities rival those of NC crows on every task given—further evidence against tool use as a uniquely human trait. In the words of Bird and Emery,

> Our results contradict suggestions that tool use was the driving force behind the evolution of advanced physical intelligence. It appears more likely that corvid tool use is a useful by-product of a domain-general "cognitive tool-kit" rather than a domain-specific ability that evolved to solve tool related problems. Whether or not each species taps into this capacity for tool use may depend on their ecology. (10374)

These findings challenge the notion that tools somehow make humans more human by transforming their cognitive abilities, and they begin to disavow the sweeping claims made by technologists and historians about humans' exclusive access to technology and about the links between development of technology and development of human cognition. Kranzberg and Pursell claimed in 1967 that "Modern physiology, psychology, evolutionary biology, and anthropology all combine to demonstrate that *Homo sapiens* cannot be distinguished from *Homo faber,* Man the Maker. We now realize that man could not have become a thinker had he not at the same time been a maker. Man made tools; but tools made man as well" (Kranzberg and Pursell 1967, 8). But the corvid studies imply otherwise.

Humans made tools, but we are not the only species capable of doing so—it seems that technological or tool-ish reasoning is possible for several species, even those without native tool experience. Given the right environmental conditions and situations, at least one other species, corvids, can fashion and employ tools—and other species may have the capacity to do so, even in the absence of actual tool use in the wild. Taking the rook studies at face value, we can no longer consider our ability to use tools as fundamental to our cognition.

WESTERN SCRUB-JAYS HAVE THE ABILITY TO PLAN

Another member of the corvid family of birds, the western scrub-jay, has shown the ability to plan for its future, something that was once considered the domain of humans. A growing body of research, much of it led by or involving Nicky Clayton, an experimental psychologist at the University of California at Davis, points to scrub-jays' ability to plan for the future, anticipating future needs and recalling the past (Wohlforth 2010). Clayton's finding have revealed that "scrub-jays plan for the future, recall incidents from the past, and mentally model the thinking of their peers" (46). It seems that in caching food for later consumption, scrub-jays have been found to dupe other birds: if they are being watched, they cache food in one place and then later move the food to a safer location (47). Clayton's studies are conducted mostly within the scrub-jays' own enclosures, because scrub-jays prefer to cache in their own territory (46). Clayton group experiments have demonstrated that jays can plan for the future in sophisticated ways:

> Given the opportunity in the evening to place a cache in either of two cages—one in which they had previously been hungry at breakfast time and one in which they had previously been fed—the birds made the correct choice, without practice, provisioning the cage where breakfast had not been provided in the past. (46–47)

According to the researchers, this type of recall demonstrates jays' ability to perform "mental time travel"—they can remember the past and plan for a future based on that past. According to Correia et al. (2007), "many people have assumed that nonhuman animals were cognitively stuck in time, incapable of acting on the basis of either the recollection of specific past episodes (retrospective cognition) or the contemplation of possible states of affair beyond the immediate future (prospective cognition" (1). But scrub-jays demonstrate that they can "provision for a future motivational state" and can indeed cache food with regard to their future preferences—behavior that perhaps indicates prospective cognition (1); and their caching of food in areas where they were once hungry also indicates a type of retrospective cognition. Researchers in one study also showed that scrub-jays "can dissociate their current and future motivational states" (2). In this study, birds were allowed to cache a number of types of food after being pre-fed one type; their cache preferences were not affected by their pre-feeding, indicating that "the birds can anticipate and take appropriate action toward the satisfaction of a future need, one that is not currently experienced" (4). Perhaps related to their retrospective and prospective cognition, scrub-jays are complex social thinkers who are capable of deceiving others (which requires a form of mental modeling). In one of Clayton's experiments, scrub-jays "remembered if they were being watched by other birds when they cached and by which ones" (Wohlforth 2010, 47). Observations show that these birds would "wait until a potential thief was distracted . . . or would chose a spot that was dark or otherwise difficult . . . to see" (47). If another bird could possibly hear where the cache was being made, scrub-jays "would choose quieter material in which to dig—sand rather than pebbles" (47). When caching with other birds looking on, scrub-jays returned to the cache spot and relocated the items "when conditions permitted privacy" (47).

To test whether this seemingly deceptive behavior was conditioned or learned, the researchers raised a group of scrub-jays by hand, never allowing them the opportunity to intrude upon the caches of other birds; these "naive jays" were found to "not take precautions to avoid being victims of theft." They explain, "Apparently, the ability to avoid theft by others depended on projecting a bird's own experience. It took a thief to know a thief" (47). Caching behavior has also been observed in ravens, so scrub-jays are likely not unique in their abilities (47). And while this is not tool use, *per se*, these experiments showing corvids' mental capability demonstrate that these birds have the mental capacities necessary to appreciate technology—the ability to plan for future use and imagine potential states of affairs. There are important components to technological knowledge that require this sort of thinking.

EXCITING CONCLUSIONS

The state of the current research indicates that NC crows are able to craft and use a variety of tools, even using metatools on occasion to do so—abilities that were once considered the hallmark of humanity (or at least of hominidity). The research on other corvids—rooks and scrub-jays—also indicates that birds have a greater cognitive sophistication, demonstrated in their use of tools, than was previously suspected; the rook research also points to the importance of environment in the use and expression of capacities for tool use. While rooks may not use tools in their native environments, the *capacity* for tool use among these birds was a very surprising finding, one that indicates the need for more research on the subject.

In my next chapter, I present contrast cases to the ape, cetacean, and corvid cases we have already examined. These contrast cases will help to flesh out a spectrum of animal technological behaviors, some more intentional and reasoned than others. I do not claim that all animals have the capacity for tool use or technological behavior, or that all seeming cases of animal tool use are equally complex or cognitively demanding; in fact, I present some cases that seem on the surface to be examples of animal tool use, but that instead stem from instinct or express the animal's extended phenotype. I also examine whether there is a significant difference between animal tool use and other construction behaviors that are very commonly found across the animal kingdom, such as the building of "constructions" (nests and dams and webs: constructions are made objects that do not have to be detached for use). I do not find that there are significant differences between construction and tool use; although tool use has often been framed as something that is critical to defining "higher intelligence," some constructions require just as much learning and crafting as tool use. It is better to think of technological thinking and knowledge along a spectrum, rather than as a system of binaries.

NOTES

1. Most of the qualities that have been proposed as condensing or defining the moral difference between humans and non-humans cannot hold either. Whether associated with tool use or not, planning, rational thought, calculation, manufacturing, empathy: none of these things can be said to separate humanity from other animals. We now have to talk in degrees, and of types of creatures, rather than some straightforward dividing line.

2. The authors note that capuchin monkeys have been able to unbend wire to get at honey, so not all primate studies are so disappointing.

3. I discovered by emailing one of the researchers that this is the same female crow as the one from the Weir et al. 2002 study and the Chappell and Kacelnik 2002 study. Her name is Betty. The male crow, named Abel, refuses to participate in tasks when separated from his female friend, so he was left out of this first experiment. He wasn't keen during the Weir et al. 2006 study either. These crow researchers started naming their crows in papers in 2006. The practice of naming individual animals in research has been a source of controversy in ape studies, and it is only recently that the practice became common in these bird studies. Thanks to Jackie Chappell for so kindly responding to my emails on this matter (Personal Communication 2010).

4. Both imitation and trial-and-error play roles in human learning, let's not forget.

5. There would have been more trials, but Betty died (Weir and Kacelnik 2006, 329).

6. Note that the researchers talk about "innovative behavior"; they also use the term "tool-oriented behavior" (331 and 318).

7. Later in this book, I will examine how evolutionary history and environment affect inter-species comparisons like these. These comparisons are often not useful, in part because the capacities of one species are used as the standard by which other species' capacities are judged. This is perhaps most problematic in the human standard that we set up for intelligence, because our particular form of intelligences comes from our specific evolutionary history. (We do not only use this intelligence standard to compare animals' capacities with ours; we also compare the intelligence of culturally and economically privileged humans with that of less privileged humans, a practice that leads to oppression, exploitation, slavery, genocide, etc. I will return to this topic later.)

8. The experiment is described by the lead author in this YouTube video that has been produced for more general audiences: https://www.youtube.com/watch?v=ZnqUAsyOTv4.

SEVEN

Spiderwebs, Beaver Dams, and Other Contrast Cases

In a study that I like to think of as "Trading Spaces: Beaver Edition" (after the now-defunct TLC television program where people redesigned rooms in their neighbors' houses), pairs of beavers with known histories of dam-building activities were moved to new environments.[1] The researcher found that these beavers moved to ponds and large rivers ended up burrowing into the banks of the shore and set up their beaver housekeeping there, where they demonstrated no interest in building dams or behaviors related to dam construction; beavers who were moved to areas with streams, however, engaged in dam-building activities, constructing their dams at thinnest part of the stream (Jung 2007). In phase two of this study, the researcher recorded the sound of water rushing and played the recording through speakers in areas known to have beaver. The recording was played during one night; the researcher returned to find "the speakers buried under several feet of sticks, gravel and mud—thus effectively silencing the sound" (np). This same procedure, repeated at several locations—along dammed streams, large rivers, lakes, and ponds—produced the same result: beavers always muffled the speakers to get rid of the sound of rushing water (np). This study demonstrates "why beavers always pick the narrowest and most shallow section of stream to build their dams—it's because that's where the noise is" (np). Beavers do *not* seem to set out with an idea to build a dam because they carefully consider its usefulness or plan out where the best place for a pool might be; beavers are compelled to literally bury the sound of rushing water, and this produces their dam-building activity at the narrowest, loudest part of the stream. Dams come not from some well-intentioned goal or plan,[2] but from environmental irritation or trigger that cues the construction behavior.

Despite these insights, beavers' dam-building activities are often pointed to as examples of excellent engineering (inspiring several universities to take up the beaver as their mascot: MIT, Oregon State, American River College). Beavers have been known to transform ecosystems by changing waterways and felling trees, sometimes destroying wildlife areas. In a study on the hydrogeomorphological effects of dam-building, one geographer explains:

> A characteristic of beaver ecology is their ability to build dams, and, thus, to modify the landscape to increase its suitability for their occupation. This ability to modify their habitat gives beaver enormous significance as geomorphic agents and has caused them to be described as "ecosystem engineers." The consequent direct and significant control on ecosystem structure and dynamics has led them to be considered a "keystone species." (Gurnell 1998)

Beaver dam-building appears intelligent: beavers always construct their dams at the narrowest part of streams, which requires fewer materials and less effort than building at wider stretches would. These dams help beavers exploit the stream by making food-finding easier. Dam-building therefore is often framed "as proof of the beaver's intelligence and engineering skill" (Jung 2007). However, observational and behavioral studies of beavers suggest that dams are built where they are because of noise volume, not because of ingenuity. Beavers need not have an understanding of folk physics to build their dams.

Beavers' building activity differs greatly from dolphins' hunting techniques and New Caledonian crows' use of tools. Like beavers, spiders also display instinctive, environmentally-cued, or compulsive building behaviors. The spinning of webs involves the production of silk threads that come from the animal's body and alter the spider's external environment to make it more useful or productive for the spider. All spiders can produce silk threads, but only web spiders built webs (Foelix 1996). Some speculate that silk threads were first used for tripping prey and were then enlarged over the spider's evolutionary history to make larger living and dining spaces (148). Orb spiders, who build highly complex webs, are considered "the evolutionary summit of web-building spiders" (145). Fossil evidence indicates that the orb web was first woven about 100 million years ago (145), but the web form has not seen any significant changes or improvements in quite some time. Spiders subjected to drug experiments have created some unusual webs, but these new productions cannot be said to be the natural product of the spider,[3] nor should these varying webs imply ingenuity, creativity, or intentionality; indeed, many of these webs are likely less effective than the normal orb web.

Given what we know about the lives of spiders and their young, there seems to be no possibility that spiders' web-building is the product of a socially transmitted material culture or of social learning of technique;

web-building know-how could not be passed along by anything other than genetics, as spiderlings spend no time with their parents or in large groups. In what *looks* like a giant collaborative project (and was reported as such in popular news media), millions of spiders and spiderlings near Lake Tawakoni in Texas produced silk threads that likely rode air currents to produce a ridiculously large spider web that covered a large area of trees; researchers on the scene identified at least eleven different spider groups involved in the production of the massive web (*ScienceDaily* 2007). However, researchers trace the giant web not to intentional planning but to a particularly good year for spiders' survival. Wet summer conditions produced many small insects for the spiders to prey upon; this abundance of prey and the favorable environmental conditions produced a great population boom for the local spiders, whose silk threads converged into one giant web.[4]

I've already introduced the idea that technology is not an all-or-none phenomenon. Rather, technological behaviors exist on a spectrum, and defining what is and is not considered "technology" is not always an easy task. Spiders spin webs, beavers build dams, and bees dance, but these constructions and techniques do not indicate technological knowledge; in this chapter, I provide cases of animal tool use that may *not* count as technology in the way that humans, New Caledonian crows, dolphins, or chimps produce and use it. Humans, apes, cetaceans, and corvids display flexibility in language and in their technical maneuvering of objects, and they also spend extended periods bonding as juvenile-adult pairs; the technical products (tools and techniques) made by these animals reflect a standardization, an ingenuity, and a complexity of cognition (understanding) that does not appear in a host of other animal tool-use behaviors. However, cases of animal construction—beaver dams and spider webs—can still appear on the same spectrum of behaviors, and we can examine the technical skills needed to produce the constructions and what knowledge is instantiated in them (more on this in the next chapter). I offer contrast cases here to help flesh out what a spectrum would look like with regard to animal artefacts and knowledge.

THE EXTENDED PHENOTYPE

Richard Dawkins provides an analysis of spider webs and beaver dams in his *The Extended Phenotype* (1982). A phenotype is the set of characteristics of an organism, and it springs from the organism's genes and environment. The *extended* phenotype, for Dawkins, is "all effects of a gene upon the world" (293): to Dawkins, beaver dams and spider webs are parts of beavers' and spiders' extended phenotypes, for they extend the animals' genes into the world. Dawkins explains that:

> an animal artefact, like any other phenotypic product whose variation is influenced by a gene, can be regarded as a phenotypic tool by which that gene could potentially lever itself into the next generation. . . . It is as though some species have shifted the burden of adaptation from bodily phenotype to extended phenotype. (199–200)

A beaver dam enables a beaver (or beavers, since they often build in mated pairs) to travel more easily by water in the newly made pond, which allows them to move wood and to forage for food with a lower risk of being hunted (200). Dawkins thus asserts that *the entire lake* may be seen as part of the beaver's extended phenotype, "extending the foraging range of the beaver in a way which is somewhat analogous to the web of a spider" (200). Spiders prove to be even more straightforward cases, because they build webs only as individuals. Dawkins says that, despite knowing of no genetic reason for spiders' web-building behaviors,

> individual spiders have consistent idiosyncrasies which are repeated in web after web. One female Zygiella-x-notata, for instance, was seen to build more than 100 webs, all lacking a particular concentric ring. Nobody familiar with the literature on behaviour genetics would be surprised if the observed idiosyncrasies of individual spider turned out to have a genetic basis. Indeed, our belief that spiders' webs have evolved their efficient shape through natural selection necessarily commits us to the belief that, at least in the past, web variation must have been under genetic influence. (198–99)

Dawkins points to a genetic origin for some spiders' consistent building of webs flawed in the exact same ways, just as there is a genetic influence that explains the standardization of orb webs among orb web builders and of silk thread among spiders that do not build webs. Spider web construction seems to be determined by genetics.

Researchers of animal tool behavior have worried that this type of genetic influence might be driving tool use among dolphins and crows, and to rule out this possibility, crow researchers have attempted to document the ways crows increasingly improve tool forms (which would indicate that the crow learned to make the tool and improved their skills over time). Similarly, dolphin researchers have tried to rule out genetic causes for dolphins' tool-using foraging techniques and hunting styles. This task is made more difficult because parents often transmit knowledge of techniques to their offspring socially or culturally, and parents and their young share a genetic resemblance. But the claim that dolphins actually learn techniques (rather than performing behaviors out of genetic influence) is bolstered by the fact that they can learn new behaviors in captivity from trainers, who share little genetic material with dolphins. Nothing similar can be said for orb web construction—spiderlings do not learn, whether by trial and error, social modeling, or teaching.

OTHER CASES

Animal Tool Behavior (Schumaker et al. 2011) documents a wide range of cases of animal tool use, tool shaping, and tool making. Here I provide brief sketches of some relevant cases.

Tool use among octopuses was first reported in 1963, when researchers saw octopuses use stones "to prop open the shells of large bivalves" to eat them (Schumaker et al. 2011). More recently, octopuses have been found to use coconuts as defensive tools; the authors specifically refer to this as tool use and talk about the possibility that invertebrates are using tools (Finn et al. 2009, R1069). They observed "soft-sediment dwelling octopuses carrying around coconut shell halves, assembling them as a shelter only when needed" (R1069). Carrying these half-shells around requires the octopus to walk in a "novel and cumbersome way," which itself seems to be less efficient and safe than normal walking (R1069). This use of coconut shells was significant because of the dexterity and manipulation required for this maneuver.

By many metrics, this octopus behavior seems to qualify as tool use. There are several dimensions by which we might identify animal tool use. The first might be *intentionality*, or, relatedly, *anticipation or planning for the future.* I've already provided more detailed accounts of scrub-jay caching, as well as Santino's rock-hoarding; it seems that these octopuses with their coconut shells are also anticipating something. Next we might consider whether there is a *material culture* or *social learning* component to the tool use. Chimp and dolphin groups give us the clearest examples of these, but NC crows also spend time modeling tool use in juvenile-adult pairs that seems to map onto the ways that we humans pass down our knowledge of technique. Third, *variety, complexity, and standardization of tools, techniques, and tool forms* may also be important. Chimp tool kits and the vast array of NC crow tool-use behaviors, including tool making and tool standardization, seem to meet these criteria. Other animals perhaps show less diversity of tool use and making, but the *techniques* of cetaceans might also be recognized for their complexity and variety. We might also consider spider webs as complex and specialized objects, and they are certainly standardized. Fourth, we look at whether forms of tools or techniques are *improved or changed over time* and whether innovative behavior takes place. There are chimp examples of innovation (Kanzi is best known), as well as cases of crows and rooks who are able to solve problems using innovative material solutions (Betty the crow). We might also take into account *cognitive abilities* (which perhaps goes along with the first criterion, intention or planning). Much of the current research on birds and cetaceans examines their cognitive capacities, which appear to go beyond the level of cognition needed for tool use and to approach empathy and the capacity for understanding. Finally, we might take *causal reasoning* to be the most important axis of tool use: we might say that the use of tools demonstrates physical causal reasoning like no other characteristic can. Great apes, capuchin monkeys, and some birds seem to count as causal reasoners, as demonstrated in their sequential use of tools for a purpose (metatool use) and some understanding of folk physics.

These criteria and cases are broad, and some may object that they are too inclusive. Some researchers assert that simple behaviors, "such as the use of an object (or objects) as shelter[,] are not generally regarded as tool use, because the shelter is effectively in use all the time, whereas a tool provides no benefit until it is used for specific purposes" (Finn et al. 2009). This is why birds' nests and animal dens are not usually considered tools. This definitional limit would also rule out hermit crab shell use, because the "tools" are always being used functionally whenever they are being handled (or bodied, in this case).

This limitation also helps us understand why beaver dams and spider webs are not tools, although, like hermit crab shells and birds' nests, they require both production and manipulation—they are made, useful, complex, and somewhat standardized, and would seem, then, to have at least some characteristics of tools. My goal is not to defend a certain definition of tool use, but to consider these definitions and delimiting objects and behaviors alongside one another on a spectrum that accounts for all these sorts of behaviors, which I see as related in many ways. In fact, I would like to expand the spectrum of technological behavior to include construction behaviors—and there is good reason for this move.

Tool use *per se* is not remarkable; many animals do it. The problem that I hope to remedy is humans' framing of tool use and technology as a form of gatekeeping, a way to define ourselves as *not-animals*. As ecological biologists Mike Hansell and Graeme D. Ruxton (2007) have argued, the demarcation between construction behaviors and tool use is arbitrary; there is little reason to consider tool use as categorically different from other animal construction behaviors. Similarly, I argue that any good (meaningful, useful) categorical distinction between tools and technology is arbitrary. In fact, at one end of the spectrum, it might be useful to consider construction behaviors in the context of technological behavior.

CONSTRUCTION AND TOOL USE

Spider webs and beaver dams also have a place on the spectrum of technological behavior, and literature on animal architecture is relevant to many discussions of tool use and technology. The most commonly cited definition of tool use (Benjamin Beck's, which stipulates that a tool is an object that is not part of the animal itself, is not attached to the environment itself, and can be manipulated to achieve some outcome; Beck 1980), *indicates no prerequisite cognitive ability* (Hansell and Ruxton 2007, 74). Technical ability is, however, often conflated with intelligence—a side-effect of definitions of technology that include the human clause.[5] It almost seems tautological: "technology defines humanness, so stuff animals do can't count as technology," as if there's an *a priori* assumption built in that technology is closed to animals. The human clause also entails the view that tool users are more advanced, evolved, or special, like in the way people casually consider humans more fully evolved and other animals as somehow lower, which is problematic in so many ways. Some of the problems stem from the human clause activating or giving credence to particular and technocratic ways of valuing people and also privilege certain ways of being in the world, modes that value human intelligence for its productive power over nature and those deemed "other." This way of seeing the world and technology's role in it also serves colonialist narratives about civilization and power.

Animal architecture expert Mike Hansell develops an account of animal construction behavior that he thinks should be situated within a wider discussion of tool use (Hansell 2007, Hansell 2005, Hansell and Ruxton 2007). According to Hansell, the attribute shared by all animal constructions is that they "extend the control of the builder over some aspect of the environment" (Hansell 2005, 1). Definitions of technology seem to highlight this very aspect of construction behavior: *technology is somehow meant to extend one's control over one's environment or world.*

Many philosophers, thinkers, and scholars have viewed technology as humankind's extension of self into the world. José Ortega y Gasset wrote that "human beings are technical, are capable of modifying their environment to fit their sense of convenience" (quoted in Mitcham 1994, 56). And Martin Heidegger emphasizes the sense of technology as extensions of humanness: "Humanity humanizes the world, injects it, impregnates it with its own ideal substance" (quoted in Mitcham 1994, 56–7). Historians of technology Melvin Kranzberg and Carroll W. Pursell, Jr. have called technology "man's efforts to cope with his physical environment . . . and his attempts to subdue or control that environment by means of his imagination and ingenuity in the use of available resources" (1967, 4–5). Historian Thomas P. Hughes entitles one of his books *Human-Built World,* as if to emphasize humanity's construction of environments (2004), and engineer and historian Henry Petroski has entitled one of his popular books *Remaking the World,* highlighting the ways in which engineers have remade the environment (1998).

Don Ihde, in *Philosophy of Technology* (1993), explains that modification to the environment, though not uniquely human, can be extended with technological capabilities:

> Even without technologies animals, and ourselves as human animals, modify at least local environments. And lions modify the very local environment of the sand by making the funnel shaped holes into which their prey may fall just as humans, in building shelters, modify envi-

> ronments, whether minimally as with desert nomads, or maximally as with contemporary urban dwellers. . . . Technologies, however, allow this modification to be amplified or magnified. (Ihde 1993, 51)

Clearly, then, the environmental modification that seems to define technology exists on a spectrum, and beaver constructions—dam-building—modify the natural world far more extensively than, for example, human desert nomads. In fact, beavers are often the scourge of wildlife resource managers, because they wreak havoc on environments that wildlife specialists are often trying to maintain, preserve, and protect. Measuring technological behavior by the amount of environmental modification it produces seems inappropriate, though, especially if these modifications are scaled to reflect the ratio of the size of the organism to the amount of modification it produces.[6] In fact, if this were the case, some animals might rival humans in their environmental impacts, for Mike Hansell notes that the "height of a large termite mound is the equivalent of three times the height of our highest buildings" (2007, 180), and, as Richard Dawkins has argued, an extended phenotype can manifest itself for miles: "Just by talking about a few examples of animal artefacts [webs and dams], then, we have pushed the conceptual range of the gene's phenotype out to many miles" (1982, 200).

Benjamin Beck himself noted of his definition of tool use that "the distinction between tools and external construction" was "hazardously arbitrary" (quoted in Hansell and Ruxton, 2007, 73). Hansell works to reconsider animal constructions (nest building, spider webs, and beaver dams) as belonging to the same spectrum as other forms of tool use identified by Beck, like woodpecker finch stick-tool making and chimpanzees' use of tools for termite fishing (74). Both Beck and Hansell still exclude other animal behaviors: sea gull mollusk-shell dropping to crack open the mollusks, chimp fence-climbing to reach food, and elephants rubbing themselves on trees are all considered non-tool use behaviors[7] (74). Hansell offers a two-part definition of construction behavior: (1) "something must be constructed" and (2) "it must necessitate behavior" (74). By this definition, sculpting and burrowing to make nests would count as construction behaviors, as would dolphin sponging for foraging, crow and rook stick and leaf tools, spider web building, etc. Construction requires only behavior, not intentionality or learning, which are more difficult to measure. Treating construction and tool use as different behaviors, Hansell and Ruxton argue, creates two distortions. First, construction behavior is left out of discussions of the evolution of intelligence, when such information might prove relevant and helpful to these discussions. Second, construction behaviors receive less scholarly attention because of the focus on tool use; even when a species has both tool use and construction behaviors, the construction behaviors often go unexplored because of the emphasis on tool use and intelligence (74–75).

The separation of tool use from construction behaviors is therefore detrimental to evolutionary biology as a field and to the study of the evolution of intelligence (75).

Hansell and Ruxton explain that construction behaviors are often *more* complex than tool use or making:

> Consider the nest building of the red-headed weaver bird. . . . To make its nest of fresh "green" twigs, this weaver bird first trims the leaves from a twig, then breaks it away from the main stem in such a way as to include a long, flexible strap of bark projecting from the broken end. This strap is then wound round the end of a narrow tree branch so that, on drying, the twig is firmly attached. Drying bark straps are the principle by which the whole nest is held together. The bird now attaches more twigs to form a stout nest attachment hanging below the branch. Below this a globular nest chamber and a downward-directed entrance tube are added. In the light of this, we have no reason to suggest that as a generality tool use will require higher levels of cognition than other construction behaviors shown by the same or closely related taxa . . . *it is important to avoid the presumption that all construction behavior (and indeed all tool use) requires a similar level of cognitive complexity.* (75, my emphasis added)

Nest building is often thought to require little intelligence and to be a product of instinct, but for some species, like the weavers described above, "there is experimental evidence of a significant learning component to nest building" (75). For example, bowerbirds' nests are also complex constructions whose techniques are likely learned: "there is also good circumstantial evidence of a learned component to bower construction by male bowerbirds," who, as juveniles, "habitually visit the bowers of mature males and observe their behaviour over a period of five or six years" (75). Not all construction behaviors are less complex than tool use; in many cases, construction behaviors actually demonstrate a greater flexibility and more complexity than cases of tool use do. Therefore, when considering the spectrum of technological behaviors, we would be remiss not to place construction behaviors within the same realm.

We can place spider web construction, which requires know-how, on one far end of a spectrum of technological behaviors. Webs certainly fit some of the criteria that define a tool or a construction: they are external objects that help aid their makers, and they demonstrate a significant level of complexity and standardization (though this is perhaps genetically encoded). But webs fail on a number of other possible criteria: they don't seem to be the product of intentional behavior, and, as such, demonstrate no innovation, creativity, ingenuity, or simple improvement over time, and they are not the product of a material culture or social learning (unlike some nest construction behaviors).

MORE CRITICISMS OF BECK'S DEFINITION

Computer scientists Robert St. Amant and Thomas E. Horton suggest in their article "Revisiting the Definition of Animal Tool Use" that Beck's definition of tool use is inadequate, not on the grounds that it does not properly include construction behaviors, but because it fails to account for behaviors that "mediate the flow of information" (2008, 1199). Motivated by cases of dolphin sponging and gorilla tool use (using a stick to test a water's depth before wading through), St. Amant and Horton explain that these instances are popularly seen as tool use, even though Beck's definition would simply rule them out (1200). St. Amant and Horton argue that detachment—one of Beck's necessary conditions is that a tool is something that can be moved around—should not be part of the definition of tool use, although detachment may be *useful* to the tool user, and may be characteristic of most tools (1200). Instead, they say, tools' function is often "mediating information flow between organisms," and they cite, for example, the implicit communication of a brandished stick (1200).

St. Amant and Horton attack all three elements of Beck's definition: detachment, alteration of the object, and alteration of the environment. They argue that, while tools are often detachable, this is better regarded as a contingent property, and not a necessary one (1201). They give three scenarios from the bird studies literature about experimental apparatus, and note that Beck's definition would include only one of these scenarios, although they all work by the same principle. All the examples have to do with pulling on a string with either a reward or a hook to use to get a reward, so the action used by the subject of the experiment is the same, though some qualify as tool use and others do not, using Beck's definition. St. Amant and Horton explain that

> part of the problem in these two scenarios [one involves pulling up a reward bucket through a transparent tube and the other has a hook with which to grab a bucket] is deciding what constitutes an object. In simple cases, the property of being unattached from its environmental substrate is sufficient to define an object, but in other cases this poses problems, such as deciding whether some composition of materials should be treated as a single object or multiple objects. . . . [M]anipulability of an object is a more informative indicator of tool use than whether the object is unattached. (1201)

These authors give further examples: an elephant sprays water from its trunk to cool itself and a chimp cups water and splashes it on its body. They maintain that Beck would accept the elephant example as tool use but would reject the chimp example; this, they say, is counterintuitive.

St. Amant and Horton also take issue with Beck's second criterion for defining tools—the condition that an object or organism must be changed by the tool's use. There are several examples of animals changing their own state by using external objects that do not seem to be tool use; I think here of cows using sticks to preen, which arguably changes their condition. St. Amant and Horton refer to a similar example in which a chimp finds a hat and places it on his head. While this meets Beck's definitional requirements (the chimp has altered his own state), few would think of this as tool use. St. Amant and Horton therefore attempt to tighten Beck's second definitional element to rule out behaviors like this hat-monkey scenario. They explain that a "plausible resolution of this problem is to view tool use in terms of dynamic, mechanical interactions rather than alteration of general conditions" (1202). Although the hat may change the chimp's state, not enough is done to "produce dynamic interactions between the candidate tool and the environment" (1202).

The third plank of Beck's definition is that tool use must alter an environment; St. Amant and Horton describe three scenarios of vultures dropping stones—one hits the egg, successfully cracking it open; another hits the egg but the impact is not enough to crack it (a failed attempt); and a third misses the egg (1202).[8] They assert that Beck's definition would only define the first scenario as tool use, despite the fact that the physical actions—and the *intent,* if we dare read that into the situation—are the same in all three scenarios, despite the differing outcomes (1202). The authors explain that

> when Beck's definition specifies that tool use requires "the proper and effective orientation of the tool," we automatically fill in a dynamic, mechanical account. This shift is enough to capture our intuitions about these scenarios: in each case, the behaviour involves manipulating or dropping an object to generate an impact on a target . . . even if the target is missed, the goal of producing the impact is apparent. (1202)

St. Amant and Horton also hold that two factors are neglected in Beck's account of tool use (and in similar definitions proposed by Jane Goodall and John Alock): first, these definitions leave out the importance of *goals* in tool use, and, second, these definitions neglect the need for *deliberate control* over the object (1202).

Both goals and deliberate control point to intention as important to defining tool use. Even when the results of animals' actions are the same—for example, wasps pound pebbles into the ground to compact the ground and chimps use stones to pound nuts, both of which result in pounded ground—the goals and understanding of these actions can differ widely, and these differences matter (1202). Deliberate control over an object *and* goal-oriented behavior matter in deciding whether something is a tool; accidentally breaking off a branch should not count, while breaking a branch for some purpose should (1202). St. Amant and Horton suggest a continuum for describing control:

> We can see differing categorizations of these examples as falling along a continuum of control. In Beck's definition, the control component of tool use falls under a binary criterion: is the candidate tool user responsible for the orientation of the external object or not? We think that this is overly restrictive, that the degree of control an animal applies to a candidate tool object influences the categorization of a behaviour as tool use or not. As interactions move from being accidental to incidental to controlled, a behavior approaches the category of tool use. (1203).

Note that phrase "behavior approaches tool use": for St. Amant and Horton, the object counts less in their scheme of things than *how it is employed* in intentional action. It is upon these two principles—one which points to intentional behavior and the other which points to controlled action[9]—that St. Amant and Horton ground their revised definition of tool use. St. Amant and Horton's account dovetails with Hansell and Ruxton's account of construction behavior: both sets of authors reject Beck's definition, in part, on grounds that it does not reasonably accommodate groups of things that we feel it should, given other things that count as tools for Beck. St. Amant and Horton give their formal definition thus:

> Tool use is the exertion of control over a freely manipulable external object (the tool) with the goal of (1) altering the physical properties of another object, substance, surface, or medium (the target, which may be the tool user or another organism) via a dynamic mechanical interaction, or (2) mediating the flow of information between the tool user and the environment or other organisms or the environment. (1203)

Hansell and Ruxton might not concede that the tool must be freely manipulable, but otherwise they might agree with this definition: the construction behaviors in which they are interested do alter the physical properties of the environment and mediate a flow of information between the builder and the world. However, St. Amant and Horton see these behaviors as existing on a spectrum; they concede that "tool use arguably does not constitute a distinct category of behaviours" (1207). This new definition accommodates the examples of a gorilla testing a water's depth with a stick and dolphin sponging—in these examples, the environment is not transformed, but a flow of information or "sensory input" is mediated when the animals use a tool to make a determination or to feel along a surface (1204). These pushbacks against Benjamin Beck's original definition of tool use highlight the importance of both information and intention. I aim to address the categories of information and intention as they relate to constructions, tools, and technologies in the account I provide in my next chapter.

TECHNOLOGICAL KNOWLEDGE AND HOW THIS HANGS TOGETHER

In chapter 3, I described several varieties of technological knowledge. Tools and technologies often mediate a user's knowledge of a world or environment. Information alone is never enough to constitute knowledge; without proper environmental or cognitive context, information has no meaning. For instance, the information about how to build an orb web resides within the phenotype of an orb web–building spider and manifests itself in the world as an extension of the spider itself. The information is given a context—a physical space in the world—without the spider knowing that this information is made manifest through its spinning, and the spider does not design the web as we in engineering define "design" (the web's design is "programmed" in the spider, in some sense). However, a gorilla using a stick to test water's depth seems to involve an informational flow from the test object to the gorilla, providing information that may prove useful to the gorilla as she attempts to cross the water (Breuer et al. 2005). The gorilla in question ("Leah") repeatedly tested the depth with her walking stick as she walked through the pool, which indicates that this action was not accidental and was goal-oriented (e380).[10] It would be appropriate to call this stick a measuring device (since that seemed to be its function) and to call what the gorilla infers from her test of the waters knowledge.[11] We might also refer to the demonstrated use of this technique as showing Leah's know-how.

There are two large categories of technological knowledge that I have discussed—knowledge that resides in the maker/user/designer (knowing-how, know-how, or technological knowledge) and knowledge that is instantiated or encapsulated within the object ("thing knowledge"). We can consider each of these types of technological knowledge—know-how and thing knowledge—as a spectrum: some behaviors require more or less skill or learning than others, as Edwin Layton has discussed in his examination of the know-how of typists and great engineers (1974). Obviously, typists and engineers have very different sets of knowledge; they also differ in the amount of training necessary to perform their jobs expertly and the amount of skill required. Even with equal training, not everyone could become proficient at typing or designing plumbing specifications, so learning is not the only factor here; skill also plays a major role in whether someone can be said to have know-how. Some tasks require finesse. This is true in some scientific laboratory settings where some experimentalists simply have "the touch" or "the knack" with an instrument or apparatus. And there are some people who, though interested in experimental science, end up in another field because they lack a certain knack for tinkering, a certain skill set that cannot be put into words. I am tempted to make a similar claim about the specialization of dolphin foraging techniques: there are techniques, such as beaching, that some dolphins are less likely to attempt because they lack the skill and experience. However, I cannot properly interview these dolphins, so I can do no more than suggest that, like humans, animals may tend to specialize. For now, it will suit me to say simply that know-how is not merely a product of learning, but often also sometimes involves some skill.

In the context of animal cases, thing knowledge might be better described in terms of encapsulated information, since "knowledge" seems to imply a knower. In Davis Baird's explanation of thing knowledge, instrument makers and tinkerers *understand* the principles at work in their machines (even if they cannot properly describe it in propositional terms), and this knowledge is instantiated in a material form (2004). This material form—for Baird, a scientific instrument—contains knowledge of the world or theory it attempts to model. It seems less controversial to call this knowledge "information," which does not imply a knower. Indeed, Baird talks about how scientific instruments "de-skill" processes that used to require skill or training, allowing less-skilled technicians and workers to produce results or images or data without being required to understand the principles behind the operation of every part of the instrument. Indeed, many labs have the results produced by their apparatuses read off by computerized set-ups, running on code written by large instrument companies, rather than by skilled workers within the laboratories; workers no longer need the skill of reading data from an instrument. The data can be integrated and presented in a way that bypasses the need for high levels of training or skill or work.[12] These built-up experimental apparatuses, like many scientific measuring and imaging devices, offer a way to do work without knowledge. These devices still require some skill or training, but much of the work is done by the instrument, based on the information encapsulated by the device. I discussed Baird's separation of function from intention in chapter 3. In technological knowledge, intention is best captured by know-how—one cannot really *know how* to do something without a goal or outcome in mind. This fits interestingly alongside discussions of the role of knowledge in animal construction and tool use. Spider web construction cannot properly be said to stem from the spider's know-how, but we could make a claim about the information required for web construction. The information that the web-building spider harbors—the information that guides his web-building—cannot be said to have a knower, or at least the spider is not the knower; this is why it is useful to speak of information rather than knowledge.

We have seen that technical knowledge exists along a spectrum. In the next chapter, I will begin to convert the spectrum of technological knowledge into a map, using these two dimensions—know-how and encapsulated information—as axes along which we can map out technologies, constructions, and techniques; this will allow us to see more clearly the relationships between otherwise dissimilar objects.

A full-bodied approach to the epistemology of technology requires that assumptions embedded in our definitions of technology be made clear. In this chapter, I've discussed the classic definition of tool use, and have brought forward some objections and alternatives to this definition. These revised definitions might lead us to consider placing some animal artefacts and constructions on the map of technological behaviors that I am building. As Ihde says, "definitions are *not neutral*" (his emphasis, 1993, 47), and the definitions of construction, tool use, and technology offered by biologists and philosophers are assuredly not neutral. The definition of tool use remains subject to amendment, and tool use exists in relationship with construction and technology. What I offer in the next chapter is a move toward mapping animal cases and human cases along the same conceptual lines—conceptual lines imported from philosophy of technology.

NOTES

1. Which is important because beavers only build dams on streams.
2. And one might infer this simply from the *mess* that goes into constructing a dam—beavers fell trees in a helter-skelter fashion that is the bane of wildlife preservationists!
3. Insofar as natural products are those un-tampered with by humankind. I'm not committed to sustaining the natural-artificial distinction, but the drug webs of spiders would be considered unnatural in most discourse that uses these concepts.
4. ‹insert your own everything-is-bigger-in-Texas joke here›
5. Frederick Ferré (in perhaps another conflation of intelligence with tool use) defines technology simply as "practical implemented intelligence" (1995, 140). I actually like Ferre's definition—it doesn't require human agency and allows for animal cases—but I don't think the inference of intelligence is easy.
6. This is not to say that human beings don't have a very large impact on the environment. Indeed, we've crapped up the atmosphere and have caused other major environmental damages and serious modifications. However, I would still maintain that using environmental modification is inadequate as a scale since, when scaled to body-size, a few humans may cause less disruption than a few beavers. It seems as if the scale of modification may or may not be related to the sophistication, complexity, and intentionality of the modifiers.
7. And the same for cows preening on sticks.
8. What St. Amant and Horton describe relates well to literature on the failure of design in philosophy of engineering. In engineering, we might describe how a structure or design was supposed to work, and then explain how the intentional action did not lead to the intended result. In fact, more often than not, engineers are stymied in the design process, having to seek new materials or revise designs to properly meet new specifications, whether financial, physical, environmental, or even social. The intended outcome or goal of a design still works toward something, even when that goal is not achieved. We still call failed technologies "technologies." Similarly, we should call tools that fail to produce the intended result "tools."
9. "Intention" is not part of the language St. Amant and Horton use; they prefer to talk of goal-orientation.

10. One may object here on the grounds that Leah would not be able to express her goal, but there are plenty of cases where users and designers of a technology cannot explain why they do the things they do, so I'd like to dispatch this objection straightaway.

11. Once again, it doesn't matter what Leah would call it. We can talk about the printing press as a technology, even though the term "technology" did not enter the vernacular until the last century.

12. I'm thinking here of computerized, commercialized probe microscopes, contrasted against the requirements of early probe microscopies, something Davis Baird has actually written about elsewhere. See Baird and Shew 2004.

EIGHT

Human Bias and Technological Knowledge

HUMAN BIASES FOR HUMAN WORK

In the fall of 2010, National Public Radio (NPR) ran a special series on the topic of "The Human Edge."[1] While it is certainly (though perhaps trivially) true that human beings have an evolutionary advantage to surviving in our particular niche, the way in which stories like this are couched—with a kind of speciesist arrogance—strikes me as wrong.[2] *Most* animals have an edge for survival in their own particular niches; evolving in a particular environment makes you better suited for that environment. Pieces in the NPR series (2010) proffered various smug claims about humanity's superiority, as if we had pulled ourselves up the evolutionary ladder using our bootstraps: the human hand was presented as being incredibly useful for fashioning tools;[3] the human voice was lauded for its wide variety of sounds and intonations, so perfectly ideal for communication (even our closest chimp relatives cannot produce the sounds we can!); our ability to write stories was presented (and romanticized) as something unique to humans; even the evolution of tear ducts was presented as a clever, functional adaptation (since tears allowed humans to signal fear to others without alerting predators). This series delved into the evolution of a host of features that distinguish us from other animals.[4]

It occurs to me that a series of this type that replaced "human" with any subset of humans or with any professional group, there would be some outcry—or at least some heavy skepticism about the series. Over the course of human history, we've witnessed the violence that follows when one racial or ethnic group promotes itself as superior to other groups. If a group of physicists, for example, congratulated itself in a series like this, a paper might appear on "The Elitist Language of Physicists," deriding these hypothetical physicists' self-aggrandizement.[5] In fact, many professional groups do tend to center the importance of their own disciplines—the pages of *Chemical and Engineering News*, for example, frame chemistry as "the central science," and I've observed administrative professionals comparing themselves to the central nervous system in the body, relaying information and keeping everything running. Groups self-promote in order to maintain themselves—to keep members interested in the group and proud of what they do. I wonder if the same thing isn't going on with the NPR series: are we humans patting ourselves on the back (with our impressive hands!) for being human, admiring our own work and complexity in a self-congratulatory way?

Our evolutionary history brought about the traits we have as human beings. Scholars in the academic field of disability studies brilliantly talk about "bodyminds" both to signal the interconnectedness of our bodies and minds and to make it easier to talk about the experience of our lives in a more concise way.[6] The categories of body and of thought, like our bodyminds, are products of our evolutionary history.[7] Evolutionary pressures can bring about similar adaptations in two populations that are not related and that do not share evolutionary niches; for example, we see examples of similar bird-beak curvature among bird species that live in different regions and are not closely related but that have similar food sources. And recall the similarities of humans' and dolphins' brain structures, discussed in chapter 5. Even in different environments with different environmental pressures, similar structures and capabilities emerged. New Caledonian crows and rooks perform similarly on tests of intelligence and tool use in captivity, but only one group actually uses tools in the wild; why, if tools are so incredibly useful and provide such a survival advantage, would this be true?

Ann Johnson writes of technological knowledge in an engineering context something that applies here:

> technological knowledge must engage some material dimension of the world. Materials are never universally available, so a technology's desirability and availability will vary geographically. As a result, because of their physical nature, technologies often have local qualities to them. If technology is knowledge, then it is local knowledge—not necessarily narrowly local, but explicitly not universal." (Johnson 2006, 572)

New research actually suggests that the use of tools, in many cases, confers *no survival advantage* for some species, and can even be a net negative for others. Rooks do not need to use hook tools in the wild, because they do not need to extract their prey from any space that would require a hook. Their environmental circumstances are such that there is no need to fashion tools or use foraging strategies that would call for a tool set in the first place. In fact, the fashioning of tools might actually slow them down (or physically weigh them down), making them more vulnerable to predators and making it harder to quickly nab prey—negatives in terms of survival outcomes.

SOMETIMES TOOLS AREN'T USEFUL

The literature on technology and tool use often contains an assumption that tools are innately useful for their users' survival, with researchers trying to figure out how a tool or technique helps an animal perform some task and how the task confers some survival value. In other words, what does the animal (or person) *get* from the use of a specific tool? However, tools are not always useful; animals' lack of tool use may reflect this fact, rather than indicating their lack of intelligence. Hansell and Ruxton explain that

> to use a tool, the bird must hold it in its beak; thus in using a tool, the bird is sacrificing the use of the organ specifically adapted to foraging. There might be only a few contexts in which a tool is superior. Most examples of tool use do involve manipulation of the tool during use. It is therefore interesting to note that, although crows and finches provide the most numerous examples of tool use in birds, the parrots, noted for their general intelligence, provide few examples of tool use in the wild. A possible explanation for this is that parrots, with their ability to grasp objects in their feet as well as to manipulate them with their beaks, find few circumstances in which a tool would offer an added advantage. (2007, 77)

Tools *simply aren't useful* in some environmental contexts, especially given the physical capabilities and environmental pressures (competition or predators) of many species. Perhaps this is why constructions are more widespread in nature than tool use.

What is important in our own lives should not create our expectations of other species; if we expect animal constructions to look like ours, it becomes more difficult for us to see actual animal technological behaviors. This obscured spot perhaps explains our deep and robust interest in chimp behavior: we like to see the interesting things our "cousins" can do, we can understand what they are doing, and we can follow their intentions as they use a tool (Hansell 2007). But context matters, because evolutionary histories take place in context. Hansell and Ruxton argue that, unlike other construction behaviors, "tool use has had little evolutionary impact" (2007, 77). Tools "are generally not particularly important to animals, in ecological or evolutionary terms" (ibid.). These animal architecture researchers think that tool use is likely "*more special to researchers than to the animals that use them*" (77, my emphasis). Given the human fascination with tool use and technology—its importance to our humanity or our set of defining characteristics—this is not unlikely, as NPR's "The Human Edge" series indicates.

Why, then, have I chosen tool use as an avenue of investigation? The answer is because it is so important to us humans—and because we can learn more about our work in the world when we contrast it against the work, and the tools, of other species. We can better situate what it is *we* do with technologies and how *we* categorize the things we do when we examine those things alongside what other creatures in the world do. Broadening our survey from tools, strictly defined, to include other constructive behaviors and techniques allows us to more fairly consider relevant features. Our work as a species can only be properly understood in the context of other work, and animals' work is not limited by detachability or material product. Tool use, as a topic, relates well to both construction behavior and technological behavior; in fact, I doubt that I can find a good demarcation point between the two, for technological behavior (at least for humans) often aims at constructing something.

PLOTTING TECHNOLOGICAL KNOWLEDGE

My purpose here is to acknowledge a spectrum of technological behaviors—and for the remainder of this book I use that term to also encompass construction behaviors. Not all technological behaviors will, of course, involve an artifact or tool; instances of tool use are, however, the easiest to spot, since we as humans are always on the lookout for tool use and since tool use among animals has been a hot topic in recent animal studies literature. There are several things to keep in mind before examining the two dimensions of technological behavior that I propose.

First, we should *not* necessarily infer intelligence or flexibility from constructive or technological behaviors. As humans—champions of technology—we often make this leap from tool use to intelligence, but there are many examples in nature of construction behavior that are not "intelligent" or learned. As the spider's web demonstrates, building behavior can stem from a structured, inflexible process, and does not imply cognition, trial and error, or learned technique. Mike Hansell goes so far as to label one of his chapters "You Don't Need Brains to Be a Builder" (2007). However, in some instances, we may want to say that an animal exhibits intelligence in its tool use, and there are both laboratory studies and good inferences from experts doing studies of animal behaviors in the wild that might support claims about intelligence. However, not every example of technological behavior on the map will come from a creative or intelligent source. In fact, I intentionally chose some examples that do not seem to demonstrate intelligence or learned technique at all, to highlight this very contrast. All of these things can be mapped along similar lines, in the same dimensions of possibility; the differences are of degree rather than of kind.

Some animal species can be safely inferred to be intelligent because of our long association with them: years of observations of chimpanzees and gorillas, both in captivity and the wild, seem to fairly irrefutably indicate intelligence, and our long history of living with domesticated animals like dogs make it easy for us to see their intelligence: we are able to understand and respond to one another. Writer Ted Kerasote explains that,

> Those studies [on dolphin, chimp, parrot, etc. communication] have corroborated what I've felt about dogs for a long time—that they're speakers of a foreign language and, if we pay attention to their vocalizations, ocular and facial expressions, and ever-changing postures, we can translate what they're saying. Sometimes we get the translation spot-on ("I'm hungry"), sometimes we make a reasonable guess ("I'm sad"), and occasionally we have to use a figure of speech to bridge the divide between their culture and our own ("I love you so much, my heart could burst"). . . . Dog owners who hold "conversations" with their dogs will know exactly what I mean. Those who don't—as well as those who find the whole notion of conversing with a dog absurd—may want to consider that humans have shared a longer and more intimate partnership with dogs than with any other domestic animal, starting before civilization existed. (2007, 10–11)

When our evolutionary history has coincided for a long time with that of other species, like dogs and cats, we can make especially good inferences about what that animal wants or needs, or what an animal is trying to do—in part because that species' history (especially for domesticated animals) has made that animal better able to communicate to human beings about their states and intentions.[8] Because context matters to how an organism acts, it isn't always possible to judge whether some object or action is the product of learning or a shared culture; there simply isn't enough information about some of the material products and techniques of non-human animals for us to judge where something should sit on our spectrum.

Second, and corollary to the first point: we should not expect every intelligent species to make or use tools or to construct things. Though these behaviors are often (perhaps mistakenly) looked for as a hallmark of intelligence, tool use or construction behaviors on their own actually tell us very little beside the fact that the development of this particular skill or technique was useful in a given evolutionary history or environmental context. The spectra I offer simply help us compare technological behaviors along the axes of knowledge and information.

Third, a spectrum is not a ranking of which species is better at tool use (by whatever criteria one might use to judge such an odd competition). As we've seen in lab studies of rooks, rooks are surprisingly adept tool users—but their tool use is only witnessed in captivity. There may be some organisms that, unseen in the wild, use tools or demonstrate techniques in some situations or contexts, but if so, we just do not know about it. Additionally, the spectra I propose arrange only particular *instances* of construction or technique or tool use, not all the *possible* construction behaviors of a given species. For example, NC crows have many different constructions and behaviors, and are mapped several times because each instance must be placed differently on the spectrum; not every object they create encapsulates similar levels of information or requires equal levels of know-how—something that can equally be said of human-made things. Mapping individual technologies or techniques makes more sense than ranking individual species, for both animal behaviors and human technologies. Some of our tools and technologies require less know-how than others. The wedge isn't on a par with the printing press; similarly, not all printing presses require equal operational skill (some may be automated or require a less expert operator). In the same way, not all cars require the same amount of skill or know-how for driving: I would have a very difficult time driving a car with manual transmission, even though I am a smooth operator with the automatic transmission. What I lack is the proper skill or know-how to use one with a manual transmission, and, in fact, the car is not well-designed to my own body, as a body with a left leg amputation who would struggle to use two foot pedals at once. We might say something similar about the technologies and techniques of NC crows: some tools require fewer steps or less finesse to make or to use, and some tools work better for particular users. The account I offer here therefore has two dimensions; because I am interested in two types of technological knowledge—know-how and thing knowledge—this scale requires both an *x* and a *y* axis.

The x *Axis: Know-How*

Some objects do not require that an agent have the know-how to construct them, or even to understand how they work, in order to use them. These objects, which have thing knowledge, are said to "de-skill" their operators' work processes. Know-how, on the other hand, requires skill, and can be inferred from displays of learning (as in dolphins), of problem-solving, and/or an understanding of folk physics (as in many of the laboratory bird studies). Know-how is expressed via ingenuity, creativity and innovation, all of which indicate a greater understanding of both causation and of material know-how than do simple operation or construction. Although innovation may come from a happy accident, realizing that a new method is useful reflects know-how: you really know how to do something when you can find a new way of doing it. I think here of Kanzi's new stone-flaking technique. While throwing a rock at the ground hardly seems innovative, his *recognition* of the action's usefulness and his practice of the technique to improve it seem to demonstrate know-how about the process and the material, similar to human tinkering.

Causal reasoning plays a crucial role in know-how. Recognizing the differences in materials and their usefulness for the intended task is important. For example, the ways that NC crows, in both the wild and in captivity, pick out materials appropriate to each task indicates that they recognize what tool will produce the correct results. The studies of rooks and crows seem to indicate that they not only accurately judge how long tools need to be for different tasks, but also understand causal relations, as demonstrated by their use of metatools.

Material culture also plays a role in know-how. When culture or social learning transmits information about making, what is being transmitted is know-how: material culture indicates the presence of know-how that can be taught and learned. Different groups of the same chimpanzee species in different regions select different materials for the same tasks, even in similar environmental contexts and when alternatives are available; this points to the importance of the chimpanzee groups' material culture. I also think here of the foraging techniques of dolphins: though these techniques do not make up a material culture, hunting and foraging strategies are often passed down the matriline, without any clear evidence of genetic basis for these behaviors. What a dolphin learns to do is often what its mother does. Know-how thus has a cultural component, both in what it produces and how it produces it.

The y *Axis: Encapsulated Information*

The other axis of interest is provided by encapsulated information—what Davis Baird calls thing knowledge. According to Baird, thing knowledge is knowledge that is encapsulated in a technology or device—things *bear* knowledge, knowledge that may or may not have a theoretical counterpart (2002, 2004). Artifacts can be said to bear this knowledge when they "successfully accomplish" some function (2002). For Baird, functioning—whether or not it can be explained propositionally—demonstrates whether an object bears knowledge. Baird spends significant space distinguishing between things he views as encoding thing knowledge (the products of science) and natural artefacts (like spider webs and beaver dams) (2004, 142–144). I consider his objections here.

Baird's first objection to spider webs and beaver dams having thing knowledge is that animal and natural artefacts are distinguishable from human-made ones. Baird explains: "Whatever the analysis of them, we can distinguish objects of human manufacture—both linguistic and material—from other natural products of life" (2004, 142). To this I offer two objections. First, we come to distinguish these things through experience; I have seen preschoolers look at a really nice bird's nest and sweetly ask who helped the bird to make it. (In defense of preschoolers, baskets and nests do share some resemblance.) I've also seen a child pick up sticks intending to add them to a beaver lodge—which looks, on the surface, just like a big pile of sticks. Even as children, we have an easier time recognizing human-made objects, because we have more exposure to them, and we can pick out human-made things that are out of context: when we see a candy wrapper alongside a stream, we know that the candy wrapper doesn't "belong." However, our discriminatory ability does not imply that the difference between the two categories is one of encapsulated knowledge; *both* types of objects can encapsulate knowledge without affecting our ability to tell them apart. The assertion that thing knowledge is limited to human-made objects has the ring of another human clause. Second, the cases Baird gives—beaver dams and spider webs—are cases of construction that *do not* involve the sort of fashioning and learned techniques—in other words, the know-how, or knowledge—that we witness in the work of dolphins, chimpanzees, and crows. We may wish to apply categorical differences, based on complexity or the employment of intelligence or ability to improve over time, to animal- and human-made objects, but we cannot rule out the possibility that animal artefacts encapsulate information simply because they are made by animals.

Baird's second objection to animal artefacts' having thing knowledge is that bearers of thing knowledge are simpler and better justified than animal artefacts. It seems that this is an argument about degree, rather than kind; he explains that

> our material creations, through our various acts of calibration, connecting them with one another and with what we say, have a greater depth of justification than do animal phenomena. Spiderwebs are well adapted to catch flies. But there is no connection established between this approach to catching spider food and other possible and actual approaches. (2004, 143)

With perhaps the exception of the giant spider web in Texas, spider webs rarely connect with one another (or with anything a spider says, for that matter), but there are animal artefacts that receive greater "depth of justification" than webs. The fact that New Caledonian crows make several distinct, standardized forms of tools points to a connection and calibration that provides more justification—more evidence of know-how or causal reasoning—than the behaviors of a spider. These crows go back to retrieve a particularly good tool to use again, which also speaks to a connection between a task and approaches to solving it. And some chimp groups' repeated testing of their tools as they construct them might also be understood in terms of the calibration and connection that Baird insists upon. While these processes may not be as complicated or encode as much knowledge as the direct-read spectrometers, the crows' hook tools do involve various acts of calibration and a real connection with the world, as do the implements of chimpanzees. Certainly, Leah the gorilla's use of a walking stick to repeatedly test the depth of water suggests something closer akin to the measuring instruments Baird describes than either of the examples of animal constructions he dismisses do.

There are at least four types of tools, and only one matches the types of material creations Baird describes in *Thing Knowledge* (2004). The four general categories of tools, as taxonomized by a group of computer scientists in a different context, are as follows (Horton et al. 2010):

> *Effective tools,* which "produce a persistent effect on materials or the environment" and include hammers and other pounding devices, screwdrivers, and other things we generally call tools.
>
> *Instruments,* which are tools that "provide information about materials or the environment" and include things like the instruments Baird considers in his book; here I would also include Leah's measuring stick.
>
> *Constraining tools,* which are tools that "constrain or stabilize materials or the environment for further application of effective tools." This category includes many examples of metatools that help in the creation of some other tool. The authors give the examples of clamps, rulers, and wedges, all of which help in the creation of effective tools.
>
> *Demarcating tools,* which "demarcate the environment or materials" such that those materials or part of the environment can be treated differently; the authors give examples such as pencils and working surfaces.

Most of the animal cases discussed in this book have employed effective tools, with a few examples of instruments and constraining tools along the way. While Horton et al. focus on tools that fit within the classic definition, and therefore neglect a larger picture that might include constructions, these divisions help to explain Baird's claims. Although tool use is well-established, there simply are not many cases of animal use of instrumentation. In developing his theoretical schema, Baird considers only material objects produced by human beings in scientific contexts. But the material products of non-human animals, which perhaps do not encapsulate the scientific knowledge that humans' productions do, certainly encapsulate technological knowledge—or at least information, a term that does not imply intentionality. Though I borrow Baird's idea that some tools encapsulate knowledge about the world, the tool use of animals is hard to describe using his categories, which are deeply attached to scientific instrumentation. Baird does, however, offer a way to talk about tools and technologies (and constructions) and what they represent without reference to the maker or user's intentionality; using the concept of encapsulated knowledge, we can rely instead on the object's function.

PLOTTING FOR A REASON

Although Baird's thing knowledge divorces function from intentionality, his account remains unsatisfying, because an object's proper function is determined by reference to its goals—in other words, its maker's intent.[9] Because objects can detach from their context of discovery or development (like scientific theories), Baird thinks we can take a "thin" notion of function in place of a "thicker" account. He explains that

> a function is a purposeful phenomenon. But adding purposes adds problems. There are problems ascertaining purpose or intentions. Without access to a designer's mind or a design team's interactions, determining the intention behind some part of an instrument can be a difficult matter of reconstruction and interpretation. (2002)

For this reason, although Baird recognizes that functions involve intention, he chooses to champion a "thinner" notion, one that acknowledges "human intentions and purposes, but that attends to the reliability and predictability of our crafted artifact." Here my account differs from Baird's. By turning away from intentionality, Baird opens the door to accepting the very animal artefacts he would like to exclude from having thing knowledge. If we remove (or lessen) the importance of intention and consider only function, spider webs and beaver dams *should* count as thing knowledge of some sort. Although the spider doesn't *intend* to design the web as it does, the web still functions beautifully. Similarly, beaver dams are well-designed (they are always at the narrowest point of a stream) and functional (very useful for the beaver in a variety of ways). Humans' and beavers' intentions behind dam-building differ widely, but both make functional constructions that alter the world. Mapping encapsulated information (or thing knowledge) on one axis and know-how on the other allows us to describe both artifacts and artefacts in terms of both function and intention and to map very different objects alongside one another. By "encapsulated information" I mean the sort of knowledge contained in devices, tools, or constructions; by "know-how," I mean the knowledge it takes to work, use, or create a device, tool, or construction. A spider may not have very much know-how—most of the skill necessary is part of its extended phenotype—but the web its produces may encapsulate information about the world. This information is not "known" by the spider but is nonetheless employed. We would map a spider web low on the x axis (know-how) but fairly high on the y axis (encapsulated information/thing knowledge).

By setting up the map in this way, in two dimensions with two types of technological knowledge represented, we can actually unite discussions of design and function happening in both biology and engineering. A spider's web is a kind of biological design, and Baird's scientific devices are types of engineering design, but both types of designs encapsulate knowledge and produce a material artifact/artefact that *works* in a given environment. To produce and use the engineering designs requires a great deal of know-how; this know-how comprises learned skill and an understanding of causal or scientific phenomena—a type of know-how that the spider does not have. This map also allows us to discuss the objects that fall somewhere between biological and engineering design. For example, young male weavers watch other males construct their nests, learning how to weave materials to improve their own constructions. Some of the behavior may be instinctual, a function of the bird's genetic predispositions, but they learn technique, acquiring know-how about the nest-building process. Dolphins do the same with foraging techniques, although these don't result in a material product that encapsulates their knowledge. Therefore, most dolphin foraging techniques rank low on the encapsulated information axis and quite far to the right on the know-how axis.

This two-dimensional mapping approach clarifies two of the most frustrating conceptual issues around defining technology and tool use in animals: (1) understanding the relationship between two radically different conceptions of technological knowledge and (2) making sense of the relationship between biological and engineering design. This two-dimensional approach should also allow us to deal with conceptual issues within differing definitions of technology. Some definitions of technology are only going to include objects that involve some particular level of know-how and beyond (Ferré's aforementioned "practical implementation of intelligence," for instance).

By combining both types of technological knowledge, know-how and thing knowledge (encapsulated information), we can map out similarities and contrasts between constructive projects that may appear to be very different. I do not limit this map to physical objects or artifacts; it is also useful to map social technologies, like language or the court system or whale songs. Whale songs, which do not produce a physically manifested object, do encapsulate information about place and relation, and they require know-how that involves a strong social component.

There may be objections to this inclusion of social technologies and techniques, although other scholars have also argued that social technologies, which are similar to artifacts and produce similar effects, should be considered technologies (Pitt 2000, for instance). My purpose here is not to argue that they *should* be defined as technologies, but simply to put them on the map. (Any readers with strong objections to their inclusion can imagine the graph without them; this project does not succeed or fail based on their inclusion.)

While readers may dispute where individual technologies or techniques belong on the map, I hope they can understand the information about technological knowledge that it conveys. It is set up like a traditional graph: 0,0 is the point where *x* and *y* axes cross, and this is the location of objects that are not used, made, or modified in any way.[10]

The amount of encapsulated information is on the vertical axis; know-how constitutes the horizontal. Some things require more know-how to make or operate than do others: for example, early scanning probe microscopies required much more user know-how than do current computerized and commercial probe microscopes. Some things also contain more encapsulated information than others: computerized probe microscopes encapsulate more information than did the early ones, which had to be plotted by hand. We can map animal constructions on these same two axes. Weaver nests, for instance, require more finesse or know-how than other nest projects (indicated by the fact that juvenile male wavers study nest building for years before constructing their own), and the nests themselves encapsulate information about the environmental context and the specific knowledge of the weaver. The same goes for dolphin foraging and hunting techniques. Some of these techniques (we could map individual techniques along these same two axes) require practice and skill and thus demonstrate a good deal of know-how.[11] Whale songs rank similarly on the know-how axis, but encapsulate less information.[12] We often find that techniques rank low in terms of encapsulated information—there is no physical object to encapsulate the information. We

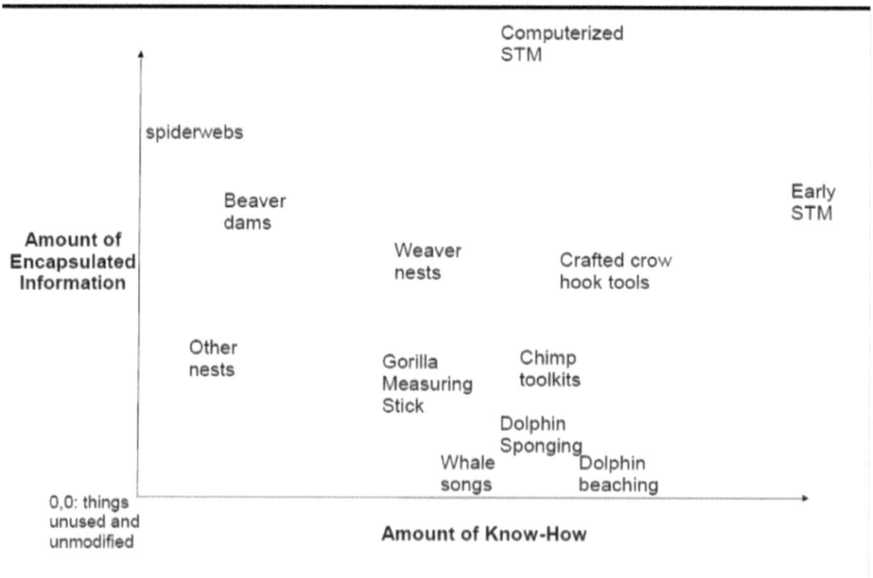

Figure 8.1. The proposed map of animal and human tools, with regard to two types of technological knowledge.
Note: STM means scanning tunneling microscope. This is one example of a very built-up human tool.

could, however (perhaps apart from discussions of technological knowledge) discuss some sort of linguistic encapsulation on the part of dolphins and whales; if language is a tool, it might make sense, for the purposes of philosophy of technology, to map this alongside material productions. The relationships I've set up on this map are not quantified in any way; they are measured only in relationship to one another. In fact, I'm not sure how one might really quantify these things. The dimensions I map help situate natural constructions alongside those of human manufacture; they allow us to consider technological knowledge in two dimensions of interest and to map the products and techniques of both humans and animals along these two axes.

CONCLUSIONS

My goal in this book has been two-fold: to argue that we should consider animal artifacts to be relevant to philosophy of technology and then to initiate this consideration in the area of epistemology. I've tried throughout this book to highlight concepts of interest in philosophy of technology, like innovation and creativity and metatool use, and especially so in discussions of the animal cases. I've also tried to stay true to how animal researchers report on their findings. I see no reason to separate the animal constructions from human projects; integrating information from animal studies into philosophy of technology is an important move toward a more wide-ranging discussion of design, function, and material culture. Rather than thinking of ourselves as a special technologically gifted species, we can see instead how our creations stem from our situation in the world, just as animals' do.

Though it once enjoyed lively debate and discussion, technological knowledge has, in the past few years, declined as an interest area in philosophy of technology. As the field has moved on from discussing how technology differs from science, *engineering* knowledge has become a much more exciting program of research for many philosophers; during the same time period, animal studies have enjoyed an unprecedented research boom in other fields. There are already movements in animal studies to discuss animal tool use and technology; animal-studies scholars are interested in what sorts of inferences we can make from our new knowledge about human nature, evolution, and behavior. While I have only addressed a small fraction of these issues, I offer a way to productively bring together discussions of the products of engineering and biology by situating them in terms of technological knowledge. Epistemology of technology offers a way to sort out different animal constructions and behaviors.

A two-dimensional graph of technological knowledge also allows for discussion between philosophers of technology about definitions. In chapter 2, I provided some excerpts from philosophers and historians of technology and from anthropologists that showed how difficult it is to define technology. I have no alternative definition of technology to defend here. If Joseph C. Pitt wishes to continue to push "humanity at work" as his definition of technology, my work does not specifically counter this. It seems to me, however, that Pitt has set the bar so that only humans can clear it.[13]

Yet the animal cases are compelling, and we need a way to discuss animal constructions; the language of epistemology of technology offers this. When Frederick Ferré gives an account of technology defining it as "the practical implementation of intelligence" (1995, 26), we can actually take out the graph and set a bar along the horizontal axis such that only those actions above the bar, which demonstrate a sufficient amount of know-how, count as practical implementation of intelligence. Spiders' webs would likely not clear this bar; some amount of built-up know-how would be necessary to implement intelligence.[14] Perhaps crows' leaf tools or chimpanzees' stick tools would clear the bar. Sociological approaches to defining technology—such as those of Don Ihde (1993) and Andrew Feenberg (1999), and perhaps also that of Joseph C. Pitt (2000)—often see social or cultural interactions as important to definitions of technology. If we leave aside for the moment the human component in these definitions, technology is, from this point of view, defined in part by social interaction, learning, and change; we can identify on our map the items that involve learned components and cultural norms. When we talk about the context of tool use in terms of "attempts to subdue or control" environments, we can do so in a way inclusive of non-human animals by using the map (Kranzberg and Pursell 1967, 5). When all creations are mapped together, we no longer need consider humans as the only source of "imagination and ingenuity in the use of available resources" (5); we can recognize all of the ingenious ways in which objects are created and employed in the world.

I'm reminded again of the satirical piece found in *The Onion* that I described at the beginning of this book. In the mock study, entitled "Study: Dolphins Not So Intelligent on Land" (2006), researchers took dolphins from their tanks, put them on land in front of some obstacle courses to navigate by echolocation, gave them tests in reading comprehension and finding land mines, and more. According to one fake researcher, "our study group offered only three types of response to every question we posed: a nonsensical, labored wheezing, an earsplitting barrage of unintelligible high-pitched shrieks, and in extreme cases, a shrill, distressed scream" ("Study: . . ." 2006).

The researchers note that the dolphins' learning curve was unimpressive—dolphins actually performed less well as the tests continued to be administered. The article ends with the following note: "many scientists believe these findings may help to explain why dolphins, for all their vaunted intelligence, have never developed technology or agriculture, or harnessed the power of fire—skills still exclusively the domain of Homo sapiens." I spent the early part of this book arguing against the inclusion of a "human clause" in our accounts of technology. We have set up a system whereby the cognitive properties of non-humans can hardly be discussed. Only in recent years has there been a relaxation of the taboo against anthropomorphism, although it seems unfair *not* to ascribe intention and other mental phenomena to non-humans in some cases. *The Onion* article amusingly highlights the ways in which we use our own experience and categories to "read" the behavior of animals such that non-humans are often set up for failure on human-made tasks. It's not clear to me that serious news specials like *The Human Edge* actually differ significantly from the *Onion* article: both begin with the assumption that there is something special about human beings that is linked to our abilities. We start with the assumption that non-human animals cannot measure up to the very standards we set up from our own abilities.

Better research methods and observation—and more extensive experience with some animals—have driven better animal studies that can make inferences about animals' capacities. Cases that surprise researchers—like Kanzi's innovative technique for stone flaking, like the complicated crafting of hook tools by New Caledonian crows, and like Leah the gorilla's measuring stick—serve to demonstrate that the *human* story that historians and philosophers have told about material culture and technology does not properly appreciate the material productions of non-human animals.

We should not mistake the story we tell about ourselves as good justification to overlook animal artefacts. By continuing to value technical ability as the most human or most intelligent expression possible, the pinnacle of evolution, we set up a system which is closed to appreciation of what animals do, as well as closed to the appreciation of others who are not "like us" in the ways we take as important. Identifying the human clause and its impacts in our conceptions of technology helps us see how our narratives about technology are biased in a way that impedes both our philosophical projects on technology and our greater appreciation of humanity in its animality.

NOTES

1. The audio recordings of this series can be found here: http://www.npr.org/templates/story/story.php?storyId=128245649.

2. I consider the sin of species arrogance a much more serious a threat to clear thinking than the threat of anthropomorphism. Mistaken anthropomorphism might make us mistakenly elevate animals in our thinking, but the species arrogance of human beings allows for much greater injustices.

3. The narrator of the piece actually says, "I have the hand of the ultimate tool maker" at one point! As someone with a body that is no longer "species typical," this

way of speaking about humans, what we do, and how we ought to be irritates me. What does it mean for my humanity if parts of my body no longer look like we expect human parts to?

4. They did at least refer to parrots' and chimps' abilities.

5. I actually saw a paper by this title delivered by Stephanie Mawler at the Brian Bertoti Graduate Student Conference at Virginia Tech (2006).

6. See the work of Margaret Price. Her book *Mad at School* (2011) is fantastic.

7. Ruth Garrett Millikan discusses this at length in her *Language, Thought, and Other Biological Categories* (1984).

8. You might be able to understand from your dog's drools and intense stare as you eat a steak that your dog wants steak. To say that you understand your dog's intentions—to get that steak by means of looking cute and/or pitiful (it has worked before)—is, I think, a reasonable inference. These sorts of inferences, of course, require experience on the part of whomever is making the inference.

9. While one might object that function can be discussed without reference to intent, to understand the purpose of a broken or new object, one must either have some account of the user or maker's intent (what is this object supposed to do?) or the context in which the object is supposedly used (where does this object belong?). There's an annex at the Smithsonian of objects whose function/purpose is unknown—Joseph C. Pitt likes to use this example in his Philosophy of Technology classes to invite debate over whether something is technology if its function is unknown.

10. Some might call these natural artefacts, but the use of the word "natural" has become suspect in philosophy of technology.

11. Breaking down all the steps in each of these techniques, and separating the learning required for each step, would enrich a graphical representation like this. I could provide a more detailed spectrum of know-how in this manner. I have, in the interest of staying on task, lumped their behaviors together; ideally, a more in-depth study of dolphin techniques could allow for a clearer picture of dolphins' technological knowledge.

12. We could argue about this. I don't think enough research has been done about what whale songs communicate to make a resounding case for this.

13. And (more damningly for Pitt's own interests), the bar would also exclude any intelligent alien non-human species that we might encounter, either in science-fiction scenarios or (possibly) in the future.

14. Keep in mind that the practical implementation of intelligence does not deny the intelligence of creatures (like rooks) that do not construct or use tools in the wild. Instead, it asserts that they are not practically implementing their intelligence in the wild.

Bibliography

Baird, Davis. 2004. *Thing Knowledge: A Philosophy of Scientific Instruments*. Berkeley: University of California Press.

———. 2002. "Thing Knowledge—Function and Truth." *Techné: Research in Philosophy and Technology 6* (2): 96–105.

Baird, Davis, and Ashley Shew. 2004. "Probing the History of Scanning Tunneling Microscopy." In *Discovering the Nanoscale,* edited by David Baird, Alfred Nordmann, and Joachim Schummer, 145–156. Amsterdam; Washington: IOS Press.

Basalla, George. 1988. *The Evolution of Technology*. New York: Cambridge University Press.

Beck, Benjamin B. 1980. *Animal Tool Behavior: The Use and Manufacture of Tools by Animals.* New York: Garland Press.

Bekoff, Marc, and Jessica Pierce. 2009. "Moral in Tooth and Claw." *The Chronicle of Higher Education,* October 18.

———. 2009. *Wild Justice: The Moral Lives of Animals.* Chicago: University of Chicago Press.

Bird, Christopher D., and Nathan J. Emery. 2009. "Insightful Problem Solving and Creative Tool Modification by Captive Nontool-using Rooks." *Proceedings of the National Academy of Sciences 106* (25): 10370–10375.

Bird, Rebecca. 1999. "Cooperation and Conflict: The Behavioral Ecology of the Sexual Division of Labor." *Evolutionary Anthropology 8* (2): 65–75.

Bluff, Lucas A., Alex A. S. Weir, Christian Rutz, Joanna H. Wimpenny, and Alex Kacelnik. 2007. "Tool-related Cognition in New Caledonian Crows." *Comparative Cognition and Behavior Reviews* 2: 1–25.

Boesch, Christophe, and Hedwige Boesch. 1990. "Tool Use and Tool Making in Wild Chimpanzees." *Folia Primatologica 54* (1–2): 86–99.

Breuer, Thomas, Mireille Ndoundou-Hockemba, and Vicki Fishlock. 2005. "First Observation of Tool Use in Wild Gorillas." *PloS Biology 3* (11): e380. Accessed October 18, 2016. doi:10.1371/journal.pbio.0030380

Carvalho, Susana, Tetsura Matsuawa, and William C. McGrew. 2011. "From Pounding to Knapping: How Chimpanzees Can Help Us to Model Hominin Lithics." In *Tool Use in Animals,* edited by Crickette M. Sanz, Josep Call, and Christophe Boesch, 225–241. Cambridge; New York: Cambridge University Press.

Chappell, Jackie, and Alex Kacelnik. 2004. "Selection of Tool Diameter by New Caledonian Crows (*Corvus moneduloides*)." *Animal Cognition 7* (2): 121–127.

———. 2002. "Tool Selectivity in a Non-mammal, the New Caledonian Crow (*Corvus moneduloides*)." *Animal Cognition 5* (2): 71–78.

CorellianScoundrel. 2008. "So, this bird walks into a convenience store and. . . . " Accessed November 8, 2010,http://www.youtube.com/watch?v=-Z2dyM2Buqs.

Correia, Sergio P. C., Anthony Dickinson, and Nicola P. Clayton. 2007. "Western Scrub-Jays Anticipate Future Needs Independently of Their Current Motivational State." *Current Biology 17* (10): 1–6.

Dawkins, Richard. 1982. *The Extended Phenotype: The Long Reach of the Gene*. New York: Oxford University Press.

De Vries, Marc J. 2003. "The Nature of Technological Knowledge: Extending Empirically Informed Studies into What Engineers Know." *Techné: Research in Philosophy and Technology 6* (3): 1–21.

Feenberg, Andrew. 1999. *Questioning Technology*. New York: Routledge.

Ferré, Frederick. 1995. [1988]. *Philosophy of Technology*. Athens: University of Georgia Press.

Fimrite, Peter. 2005 "Daring Rescue of Whale off Farallones: Humpback Nuzzled Her Saviors in Thanks after They Untangled Her from Crab Lines, Diver Says." *San Francisco Gate*, December 14. http://articles.sfgate.com/2005-12-14/news/17403910_1_humpback-crab-pots-whale.

Finn, Julian K., Tom Tregenza, and Mark D. Nordman. 2009 (15 December). "Defensive Tool Use in a Coconut-carrying Octopus." *Current Biology 19* (23): R1069–R1070.

Foelix, Rainer F. 1996. *Biology of Spiders*, 2nd ed. New York: Oxford University Press.

Forbes, Robert J. 1967. "The Beginnings of Technology and Man." In *Technology in Western Civilization*, Volume 1, edited by Melvin Kranzberg and Carroll W. Pursell, 11–26. New York: Oxford University Press.

Gazda, Stefanie K., Richard C. Connor, Robert K. Edgar, and Frank Cox. 2005. "A Division of Labour with Role Specialization in Group-hunting Bottlenose Dolphins (*Tursiops truncatus*) off Cedar Key, Florida." *Proceedings of the Royal Society of London, Series B 272* (1559): 135–140.

Griffin, Donald. 1976. *The Question of Animal Awareness: Evolutionary Continuity of Mental Experience*. New York: The Rockefeller University Press.

Gurnell, Angela M. 1998. "The Hydrogeomorphological Effects of Beaver Dam-building Activity." *Progress in Physical Geography* 22 (2): 167–189.

Hall, A. Rupert. 1978. "On Knowing and Knowing How to . . . ." *History of Technology 3*: 91–103.

Hansell, Mike. 2007. *Built by Animals: The Natural History of Animal Architecture*. New York: Oxford University Press.

———. 2005. *Animal Architecture*. New York: Oxford University Press.

Hansell, Mike, and Graeme D. Ruxton. 2007. "Setting Tool Use within the Context of Animal Construction Behavior." *Trends in Ecology and Evolution 23* (2): 73–78.

Haraway, Donna. 1989. *Primate Visions: Gender, Race, and Nature in the World of Modern Science*. New York: Routledge.

Henderson, Mark. 2009 "Chimpanzee's Plan to Attack Zoo Visitors Shows Evidence of Premediated Thought." *The Times Online*, March 10. http://www.timesonline.co.uk/tol/news/science/article5877764.ece.

Herman, Louis M., Douglas G. Richards, and James P. Wolz. 1984. "Comprehension of Sentences by Bottlenosed Dolphins." *Cognition 16* (2): 129–219.

Hobaiter, Catherine, Timothée Poisot, Klaus Zuberbühler, William Hoppitt, and Thibaud Gruber. 2014. "Social Network Analysis Shows Direct Evidence for Social Transmission of Tool Use in Wild Chimpanzees." *PLOS Biology* 12 (9): e1001960.

Horton, Thomas E., Robert St. Amant, John M. Daughtry, and Colin G. Butler. 2010. "Tool-based Direct Manipulation Environments." *Department of Computer Science Technical Report* TR-2010–23. North Carolina State University.

Hughes, Thomas P. 2004. *Human-Built World: How to Think about Technology and Culture*. Chicago: University of Chicago Press.

Hume, David. 1888. [1978]. *A Treatise of Human Nature*. New York: Oxford University Press.

Hunt, Gavin R. 2000. "Human-like, Population-level Specialization in the Manufacture of Pandanus-Tools by New Caledonian Crows (Corvus moneduloides)." Proceedings of the Royal Society of London, Series *B 267* (1441): 403–413.

———. 1996. "Manufacture and Use of Hook-Tools by New Caledonian Crows." Nature *379*: 249–251.

Hunt, Gavin R., Robb B. Rutledge, and Russell D. Gray. 2006. "The right tool for the job: what strategies do wild New Caledonian crows use?" *Animal Cognition 9* (4): 307–316.

Hunt, Gavin R., and Russell D. Gray. 2007. "Parallel Tool Industries in New Caledonian Crows." *Biology Letters 3* (2): 173–175.

———. 2006. "Tool Manufacture by New Caledonian Crows: Chipping Away at Human Uniqueness." Acta Zoologica Sinica *52* (Suppl.): 622–625.

———. 2004. "The Crafting of Hook Tools by Wild New Caledonian Crows." Proceedings of the Royal Society of London, Series *B* (Suppl.*) 271*: S88–S90.

———. 2003. "Diversification and Cumulative Evolution in New Caledonian Crow Tool Manufacture." Proceedings of the Royal Society, London *B 270* (1517): 867–874.

Idhe, Don. 1993. *Philosophy of Technology: An Introduction*. New York: Paragon House.

"information." 2003. *The American Heritage Dictionary of the English Language, 4th ed.* Accessed October 6, 2010. http://www.thefreedictionary.com/information.

Inoue-Nakamura, Noriko, and Tetsuro Matsuzawa. 1997. "Development of Stone Tool Use by Wild Chimpanzees (*Pan troglodytes*)." *Journal of Comparative Psychology 111* (2): 159–173.

Jane Goodall Institute. 2007. "Tool Making." Chimpanzee Central. Accessed November 10, 2010. http://www.janegoodall.org/chimp-central-toolmakers.

Janik, Vincent M. 2000 (25 August). "Whistle Matching in Wild Bottlenose Dolphins (*Tursiops truncatus*)." *Science 289* (5483): 1355–1357.

Janik, Vincent M., Laela S. Sayigh, and Randall S. Wells. 2006. "Signature Whistle Shape Conveys Identity Information to Bottlenose Dolphins." *Proceedings of the National Academy of Sciences 103* (21): 8293–8297.

Johnson, Ann. 2005. "Revisiting Technology as Knowledge." *Perspectives on Science 13* (2): 554–573.

Jung, Jim. 2007. [2003]. "Why Beavers Build Dams." The Waterman and Hill-Traveller's Companion, a Natural Events Almanac. http://www.naturealmanac.com/archive/beaver_dams/beaver_dams.html.

Kenward, Ben, Christian Rutz, Alex A. S. Weir, and Alex Kacelnik. 2006. "Development of Tool Use in New Caledonian Crows: Inherited Action Patterns and Social Influences." *Animal Behaviour 72* (6): 1329–1343.

Kenward, Ben, Alex A. S. Weir, Christian Rutz, and Alex Kacelnik. 2005. "Tool Manufacture by Naïve Juvenile Crows." *Nature 433*: 121.

Kerasote, Ted. 2007. *Merle's Door: Lessons from a Freethinking Dog*. Orlando: Harcourt Books.

Kranzberg, Melvin, and Carroll W. Pursell, Jr. 1967. "The Importance of Technology in Human Affairs." In *Technology in Western Civilization*, Volume 1, edited by Melvin Kranzberg and Carroll W. Pursell, 3–10. New York: Oxford University Press.

Krützen, Michael, Erik P. Willems, and Carel P. van Schaik. 2011. "Culture and Geographic Variation in Orangutan Behavior." *Current Biology 21*: 1–5.

Krützen, Michael, Janet Mann, Michael Heithaus, Richard C. Connor, Lars Lejder, and William B. Sherwin. 2005. "Cultural Transmission of Tool Use in Bottlenose Dolphins." *Proceedings of the National Academy of Sciences 102* (25): 8939–8943.

Kummer, H., and Jane Goodall. 1985. "Conditions of Innovative Behavior in Primates." *Philosophical Transactions of the Royal Society of London, Series B 308* (1135): 203–214.

Laudan, Rachel, ed. 1984. *The Nature of Technological Knowledge*. Dordrecht: D. Reidel Publishing Company.

Layton, Edwin. 1987. "Through the Looking Glass, or News from Lake Mirror Image." *Technology and Culture 28* (3): 594–607.

———. 1974. "Technology as Knowledge." *Technology and Culture 15* (1): 31–41.

———. 1971. "Mirror-Image Twins: The Communities of Science and Technology in 19th-Century America." *Technology and Culture. 12* (4): 562–580.

Lee, K. 2009. "*Homo faber*: The Unity of the History and Philosophy of Technology." In *New Waves in Philosophy of Technology*, edited by Jan Kyrre Berg Olson, Evan Selinger, and Søren Riis, 13–39. Basingstoke; New York: Palgrave Macmillan.

Leinhard, John. 2000. *The Engines of Our Ingenuity: An Engineer Looks at Technology and Culture*. New York: Oxford University Press.

Logan, Corina, Alexis J. Breen, Alex H. Taylor, Russell D. Gray, and William J. Hoppitt. 2016. "How New Caledonian Crows Solve Novel Foraging Problems and What It Means for Cumulative Culture." *Learning Behavior 44* (1): 18–28.

Mann, Janet, and Eric M. Patterson. 2013. "Tool Use by Aquatic Animals." *Philosophical Transactions of the Royal Society of London, Series B 368*: 20120424. http://dx.doi.org/10.1098/rstb.2012.0424

Marino, Lori. 2002. "Convergence of Complex Cognitive Abilities in Cetaceans and Primates." *Brain, Behavior and Evolution 59* (1–2): 21–32.

Marino, Lori, Richard C. Connor, R. Ewan Fordyce, Louis M. Herman, Patrick R. Hof, Louis Lefebvre, David Lusseau, Brenda McCowan, Esther A. Nimchinsky, Adam A. Pack, Luke Rendell, Joy S. Reidenberg, Diana Reiss, Mark D. Uhen, Estel Van der Gucht, and Hal Whitehead. 2007. "Cetaceans Have Complex Brains for Cognition." *PloS Biology 5* (5): e139.

Mawler, Stephanie. 2006. "The Elitist Language of Physicists." Paper presented at the Brian Bertoti Graduate Student Conference, Virginia Tech.

McGinn, Robert E. 1978. "What Is Technology?" *Research in Philosophy and Technology 1*:179–197.

Millikan, Ruth Garrett. 1984. *Language, Thought, and Other Biological Categories.* Cambridge: MIT Press.

Mitcham, Carl. 1994. *Thinking through Technology: The Path between Engineering and Philosophy*. Chicago: University of Chicago Press.

Mitcham, Carl, and Robert Mackey (eds.). 1972. *Philosophy and Technology*. New York: The Free Press.

Morelle, Rebecca. 2010 (April 20). "Clever Crows Can Use Three Tools." *BBC News,* April 20. http://news.bbc.co.uk/go/pr/fr/-/2/hi/science/nature/8631486.stm.

Morning Edition. 2010. "A Handy Bunch." In *The Human Edge* (Special Series), radio correspondent Christopher Joyce. National Public Radio, July 26http://www.npr.org/series/128245649/the-human-edge.

Nagel, Thomas. 1974. "What Is It Like to Be a Bat?" *The Philosophical Review 83* (4): 435–450. doi:10.2307/2183914.

Noad, Michael J., Douglas H. Cato, M. M. Bryden, Micheline- N. Jenner, and K. Curt S. Jenner. 2000. "Cultural Revolution in Whale Songs." *Nature 408* (537): 537.

Osvath, Mathias. 2009. "Spontaneous Planning for Future Stone Throwing by a Male Chimpanzee." *Current Biology 19* (5): R190–191.

Personal Communication. 2010 (July 1). Email Communication with Jackie Chappell, Lecturer in Animal Behavior, Centre for Ornithology, University of Birmingham.

Petroski, Henry. 2006. *Success through Failure: The Paradox of Design.* New York: Vintage Books.

———. 2003. *Small Things Considered*. New York: Vintage Books.

———. 1998. *Remaking the World.* New York: Vintage Books.

———. 1992. *The Evolution of Useful Things*. New York: Alfred Knopf.

———. 1985. *To Engineer Is Human: The Role of Failure in Successful Design*. New York: Vintage Books.

Pitt, Joseph C. 2000. *Thinking about Technology: Foundations of the Philosophy of Technology*. New York: Seven Bridges Press.

Povinelli, David J., Laura A. Theall, James E. Reaux, and Steve Gimbrone. 2000. "The Tool-Insertion Problem: The Question of Shape." In *Folk Physics for Apes: The Chimpanzee's Theory of How the World Works,* edited by David J. Povenelli, 173–205. Oxford: Oxford University Press.

Price, Margaret. 2011. *Mad at School: Rhetorics of Mental Disability and Academic Life.* Ann Arbor: University of Michigan Press.

Pryor, Karen, and Kenneth S. Norris, eds. 1991. *Dolphin Societies: Discoveries and Puzzles*. Berkeley: University of California Press.

Raymond, Midge. 2016. "No One Mourns an Unnamed Animal: Why Naming Animals Might Help Save Them." *Zoomorphic Magazine 6*. http://zoomorphic.net/2016/

09/no-one-mourns-an-unnamed-animal-why-naming-animals-might-help-save-them/.

Rowlands, Mark. 2013. *Running with the Pack*. Pegasus Books.

———. 2005. "Environmental Epistemology." *Ethics & the Environment 10* (2): 5–27.

Rutz, Christian, Shoko Sugasawa, Jessica E. M. van der Wal, Barbara C. Klump, and James J. H. St. Clair. 2016. "Tool Bending in New Caledonian Crows." *Royal Society Open Science*. doi:10.1098/rsos.160439.

Ryle, Gilbert. 1946. "Knowing How and Knowing That." *Proceedings of the Aristotelian Society 46*.

Sanz, Crickette M., Josep Call, and Christophe Boesch, eds. 2013. *Tool Use in Animals*. Cambridge; New York: Cambridge University Press.

Sanz, Crickette M., and David B. Morgan. 2007. "Chimpanzee Tool Technology in the Goualougo Triangle, Republic of Congo." *Journal of Human Evolution 52* (4): 420–433.

Sanz, Crickette M., Caspar Schöning, and David B. Morgan. 2009. "Chimpanzees Prey on Army Ants with Specialized Tool Set." *American Journal of Primatology 72* (1): 17–24.

Sargeant, Brooke L., Janet Mann, Per Berggren, and Michael Krützen. 2005. "Specialization and Development of Beach Hunting, a Rare Foraging Behavior, by Wild Bottlenose Dolphins (*Tursiops* sp.)." *Canadian Journal of Zoology 83* (11): 1400–1410.

Savage-Rumbaugh, Sue, and Roger Lewin. 1994. *Kanzi: The Ape at the Brink of the Human Mind*. New York: Wiley.

Scheldeman, Griet. 2009. "Orangutan Ruse Misleads Predators." *BBC News*, August 5.

Schick, Kathy D., and Nicholas Toth. 1993. *Making Silent Stones Speak: Human Evolution and the Dawn of Technology*. New York: Simon and Schuster.

Schumaker, Robert W., Kristina R. Walkup, and Benjamin Beck. 2011. *Animal Tool Behavior: The Use and Manufacture of Tools by Animals*, rev. and updated ed. Baltimore: Johns Hopkins University Press.

ScienceDaily. 2007. "Enormous Spider Web Found in Texas." *ScienceDaily*, September 13. http://www.sciencedaily.com/releases/2007/09/070912145919.htm.

Shew, Ashley. 2017. "Animals in Philosophy of Technology." In *Spaces for the Future: A Companion to Philosophy of Technology*, edited by Joseph C. Pitt and Ashley Shew, forthcoming. New York: Routledge.

———. 2008. "What Dolphins Want: Animal Intentionality and Tool Use." MA Thesis, Philosophy, Virginia Tech. http://scholar.lib.vt.edu/theses/available/etd-04172008–144940/.

———. 2007. "Spider Webs, Beaver Dams, and the Sticky Wicket: An Investigation on What Counts as Technology and What Counts as Knowledge." MS Thesis, Science and Technology Studies, Virginia Tech. http://scholar.lib.vt.edu/theses/available/etd-05152007-135849/.

Silva, Francisco J., and Kathleen M. Silva. 2010. "How Do Adult Humans Compare with New Caledonian Crows in Tool Selectivity?" *Learning & Behavior 38* (1): 87–95.

St. Amant, Robert, and Thomas E. Horton. 2008. "Revisiting the Definition of Animal Tool Use." *Animal Behavior 75* (4): 1199–1208.

"Study: Dolphins Not So Intelligent on Land." 2006. *The Onion 42* (7), February 15. http://www.theonion.com/content/node/45360/print/.

Taylor, Alex H., Douglas Elliffe, Gavin R. Hunt, and Russell D. Gray. 2010. "Complex Cognition and Behavioural Innovation in New Caledonian Crows." *Proceedings of the Royal Society of London, Series B 277* (1694): 2637–2643

Taylor, Alex H., Gavin R. Hunt, Jennifer C. Holzhaider, and Russell D. Gray. 2007. "Spontaneous Metatool Use by New Caledonian Crows." *Current Biology 17* (17): 1504–1507.

Taylor, Alex H., Gavin R. Hunt, Felix S. Medina, and Russell D. Gray. 2009. "Do New Caledonian Crows Solve Physical Problems through Causal Reasoning?" *Proceedings of the Royal Society of London, Series B 276* (1655): 247–254.

Taylor, Alex H., Rachael Miller, and Russell D. Gray. 2012. "New Caledonian Crows Reason about Hidden Causal Agents." Proceedings off the National Academy of Sciences of the United States of America 109 (40): 16389–16391.

Tenner, Edward. 1997. *Why Things Bite Back: New Technology and the Revenge Effect.* London: Fourth Estate.

Tyack, Peter L. 2000 (25 August). "Animal Behavior: Dolphins Whistle a Signature Tune." *Science 289* (5483): 1310–1311.

Vincenti, Walter. 1990. *What Engineers Know and How They Know It*. Baltimore: John Hopkins University Press.

Visalberghi, Elisabetta, Dorothy M. Fragaszy, and Sue Savage-Rumbaugh. 1995. "Performance in a Tool-Using Task by Common Chimpanzees (*Pan troglodytes*), Bonobos (*Pan paniscus*), an Orangutan (*Pongo pygmaeus*), and Capuchin Monkeys (*Cebus apella*)." *Journal of Comparative Psychology 109* (1): 52–60.

von Bayern, Auguste M. P., Robert J. P. Heathcote, Christian Rutz, and Alex Kacelnik. 2009. "The Role of Experience in Problem Solving and Innovative Tool Use in Crows." *Current Biology 19* (22): 1965–1968.

Weir, Alex A. S., Jackie Chappell, and Alex Kacelnik. 2002. "Shaping of Hooks in New Caledonian Crows." *Science 297* (5583): 981.

Weir, Alex A. S., and Alex Kacelnik. 2006. "A New Caledonian Crow (*Corvus moneduloides*) Creatively Re-designs Tools by Bending or Unbending Aluminum Strips." *Animal Cognition 9* (4): 317–334.

Westergaard, Gregory Charles. 1995. "The Stone-Tool Technology of Capuchin Monkeys: Possible Implications for the Evolution of Symbolic Communication in Hominids." *World Archaeology 27* (1): 1–9.

Wohlforth, Charles. 2010. "Who Are You Calling Bird Brain?" *Discover Magazine,* March: 45–49.

Yamamoto, Shinya, Tatyana Humle, and Masayuki Tanaka. 2013. "Basis for Cumulative Cultural Evolution in Chimpanzees: Social Learning of a More Efficient Tool-Use Technique." *PLoS One 8* (1): e55768.

Index

About the Author

Ashley Shew resides in beautiful Blacksburg, Virginia, where she is assistant professor in the department of Science, Technology, and Society at Virginia Tech. Philosophy of technology invigorates her thinking and provides a supportive scholarly community. Shew's research interests lie at the intersections of philosophy of technology with animal studies, disability studies, ethics and emerging technologies, and instrumentation. She serves as a board member in the Society for Philosophy and Technology and is a co-editor of *Spaces for the Future: A Companion to Philosophy of Technology* (2017). Her current research centers on techno-optimism and techno-ableism in narratives about disability and technology; she keeps a website about teaching on topics in Technology and Disability at http://techanddisability.com.

www.ingramcontent.com/pod-product-compliance
Ingram Content Group UK Ltd.
Pitfield, Milton Keynes, MK11 3LW, UK
UKHW041122071026
15587UKWH00009B/93

9 781498 543118